Environmental Justice and Land Use Conflict

Conflict over the extraction of coal and gas resources has rapidly escalated in communities throughout the world. Using an environmental justice lens, this multidisciplinary book explores cases of land use conflict through the lived experiences of communities grappling with such disputes.

Drawing on theories of justice and fairness in environmental decision making, it demonstrates how such land use conflicts concerning resource use can become entrenched social problems, resistant to policy and legal intervention. The author presents three case studies from New South Wales in Australia and Pennsylvania in the US of conflict concerning coal, coal gas and shale gas development. It shows how conflict has escalated in each case, exploring access to justice in land use decision making processes from the perspective of the communities at the heart of these disputes. Weaknesses in contemporary policy and regulatory frameworks, including ineffective opportunities for public participation and a lack of community recognition in land use decision making processes, are explored.

The book concludes with an examination of possible procedural and institutional reforms to improve access to environmental justice and better manage cases of land use conflict. Overall, the volume links the philosophies of environmental justice with rich case study findings, offering readers further insight into both the theory and practice of land use decision making.

Amanda Kennedy is an Associate Professor and the Deputy Director of the Australian Centre for Agriculture and Law at the University of New England, Australia.

Earthscan Studies in Natural Resource Management

For more information on books in the Earthscan Studies in Natural Resource Management series, please visit the series page on the Routledge website: http://www.routledge.com/books/series/ECNRM/

Environmental Justice and Land Use Conflict

The Governance of Mineral and Gas Resource Development

Amanda Kennedy

First published 2017 by Routledge

2 Park Square, Milton Park, Abingdon, Oxfordshire OX14 4RN

52 Vanderbilt Avenue, New York, NY 10017

Routledge is an imprint of the Taylor & Francis Group, an informa business

First issued in paperback 2019

British Library Cataloguing-in-Publication Data
A catalogue record for this book is available from the British Library

Library of Congress Cataloging-in-Publication Data
A catalog record for this book has been requested

ISBN: 978-1-138-88856-2 (hbk)
ISBN: 978-0-367-33531-1 (pbk)

Typeset in Goudy
by Apex CoVantage, LLC

Contents

Figures and tables

Figures

Tables

Preface

This book is concerned with conflict – in particular, land use conflict associated with natural resource extraction. Conflict over extractive resource development – such as coal and unconventional natural gas – is by no means a new phenomenon. However, in recent times such conflicts have transformed in both complexity and intensity, particularly in the rural context where extractive development is more likely to occur. Changing global energy demands, increasing economic pressures and the availability of new technologies have witnessed a rush to extract and develop mineral and gas resources. As the scale of extractive activity has rapidly expanded, previously diffuse resource users have been forced to operate in much closer proximity to one another. As a result, the impacts of extractive development have become more visible, and more profound. Some communities have actively campaigned against extractive development, expressing concerns about social, environmental and human health impacts, and scepticism of the promised economic benefits. Disputes over the extraction of natural resources seem to have become more complicated and confrontational, and in some communities are firmly entrenched as intractable social problems. Governments have been faced with the dilemma of responding to such conflict, but initiatives to date – including regulatory and policy reform – have made only limited inroads into these difficult conundrums. As the growing international appetite for energy escalates, and natural resources become increasingly contested, it is critical that the governance mechanisms implemented to manage land use conflict are effective.

Why do these conflicts elude resolution? This was the question I set out to answer when I began exploring this area several years ago. I became interested in this topic as conflict had arisen in my region (north-west New South Wales, Australia) over proposals to explore and develop coal and gas resources. Emotions ran high as communities grappled with the potential influx of development and the benefits and opportunities it would bring, along with possible threats to the environment and impacts upon the region's strong agricultural industry. Protests, blockades and civil disobedience increased as citizens sought to vocalise their anger with approvals for exploration and development, as well as express their distrust of mining and gas companies and the government. Local and national media became saturated with stories of acrimonious disputes, landholder distress and community fragmentation. Government efforts to quell conflict seemed to have little impact and, for the most part, exacerbated tensions further.

As a legal researcher, what interested me most was the role of the law in resolving these disputes, and why – despite legal and policy initiatives to better 'balance' competing interests – opposition to extractive development often persisted. It became clear that at the heart of many of these conflicts were claims about justice – but these were not simply assertions about the unfair distribution of the benefits and burdens associated with resource development. There were also concerns about whether communities and individuals impacted by resource extraction had been meaningfully included in decision making, how they were treated in those various processes and, ultimately, whether decision makers had adequately accounted for impacts upon community well-being and their 'way of life'. Their claims for justice were shaped by perceptions of *injustice*, and beliefs about what was right and fair with respect to land use decision making practice. As I delved further into the research, it became evident that there was a significant disconnect between the claims for justice advanced by communities grappling with extractive development, and the governance responses intended to deal with their concerns.

Conflict is not necessarily a binary struggle between a just option and an unjust option; there are different ways of framing justice, and conflict can arise when one conception of what is fair diverges from or contradicts another. For example, one particular framing of a problem (and potential solutions), such as the economic and energy security imperatives of pursuing unconventional gas development for the 'greater good', may not be aligned with perspectives which see justice as avoiding local environmental harm. Conflict also arises where one party misunderstands or fails to respond to another's claim for justice, or where certain perspectives dominate others and exclude alternative conceptions of what is just. Governance interventions designed to manage intractable land use conflict situations often underestimate or overlook the potential for such value-laden clashes between participants. Confronted with difficult decisions in rapidly changing circumstances, governments and policy makers frequently opt for the most expedient solution to resolve these rivalries, but this can further exacerbate conflict if fundamental justice concerns are neglected.

As my research progressed, I wondered whether governance approaches could better account for these divergent interests and perspectives. Research has shown that by focusing on claims for justice, we can gain deeper insight into how different people interpret and react to specific issues, and better understand the dynamics of conflict. This can then lead to the development of governance mechanisms that are sensitive to the wide variety of justice claims that may be advanced, and thus more attuned to situations of conflict. When the broad spectrum of issues arising in a conflict situation are thoroughly considered, decisions are much more likely to be well informed and fair, and more likely to be accepted by those affected.

The concept of 'environmental justice', which identifies specific justice issues relevant to environmental matters (including resource allocation and land use decisions), is a useful place to begin to develop this understanding. The concept conceives of justice broadly, incorporating the distribution of environmental risks and harms, opportunities for public participation, recognition of those impacted and the overall functioning of communities. Environmental justice provides a framework to explore how individuals and communities articulate perceptions of fairness and justice within a land use conflict situation, and to understand how difficult decision making conundrums may be better managed by meaningfully accounting for these perspectives. Understanding how and why justice matters is central to environmental justice; it assists in unravelling and unpacking what is fundamentally at stake within resource use conflicts.

In this book, I take these theoretical ideas about environmental justice and use them as a starting point to examine real-world case studies of conflict concerning extractive development. My aim is to understand, using an evidence-based in-depth qualitative case study approach, how perceptions of justice influence these disputes, and to explore whether a more systematic knowledge of environmental justice could be better integrated in land use decision making practice to channel conflict differently. From the perspective of communities and other actors involved in each case scenario, I explore how divergent notions of justice triggered conflict, and whether governance responses to justice claims were successful, or further exacerbated conflict. The case studies are ultimately a vehicle to explore how environmental justice concerns can be better incorporated into land use decision making processes, and towards the end of the book I discuss possible procedural, institutional and community engagement reforms in order to better manage conflict over extractive development.

In Chapter 1, I first set the scene for this investigation by explaining how placing environmental justice as the central element in land use conflict can assist in illuminating the nature of conflict over land use development associated with resource extraction. I then detail how I carried out the task of investigating real-world case studies of land use conflict from an environmental justice perspective.

There are many ways injustice may be experienced, as well as a variety of causal explanations for the occurrence of injustice. In this study, I adopt Schlosberg's (2007) broad and plural definition of environmental justice, and use the framework proposed by Walker (2012) to deconstruct claims for environmental justice. Walker's framework proposes that environmental justice claim-making is made up of three elements: justice, evidence and process. In Chapter 2, I, and my colleagues Lisa de Kleyn and Matthew Ryan, discuss further the 'justice' and 'process' elements from this framework that are central to understanding claims for environmental justice.

Chapters 3, 4, 5 and 6 turn to the element of 'evidence' of environmental injustice through an exploration of three case studies involving conflict over the development of extractive resources. Two of these case studies are located in the state of New South Wales (NSW), Australia: the expansion of the Mount Thorley–Warkworth coal mine near Bulga, in the predominantly rural Upper Hunter Valley region; and extractive developments located in the agricultural Namoi catchment region in north-west NSW (split into two parts over chapters 4 and 5): the development of coal mines on the Liverpool Plains, and the exploration of coal seam gas in Narrabri. The final case study is situated in the US, and is concerned with Pennsylvania's state-wide legislative amendments to control shale gas development, which occurs largely in rural communities throughout the state. I chose these case studies because they provide examples of situations in which conflict over planned or approved extractive development had resulted in specific governance interventions (including policy and regulatory reform). In describing the cases, I detail, using the voices of some of the key stakeholders involved, how justice was articulated and how injustice was experienced. Note that the case studies concern extractive development in rural communities within developed nations (that is, Australia and the US). Conflict over resource extraction in developing countries has been studied extensively within a justice framework (e.g. Urkidi and Walter 2011; Özkaynak et al. 2012; Hilson 2012; Coni-Zimmer, Flohr and Jacobs 2016; Tschakert 2009; Martinez-Alier 2001); whereas new conflicts that have arisen in developed countries – largely in rural areas – have not received as much scholarly attention. Nonetheless, some of the findings from this study are relevant to a range of contexts.

In the final chapter, Chapter 7, I bring together the environmental justice themes which emerge from the case studies. The studies and examples I use from the literature indicate that efforts to incorporate justice-based approaches within land use decision making practice can lead to decisions that are well informed and more likely to be fair, and ultimately accepted by those concerned. The book closes with a discussion of options for implementing justice-based governance approaches as a means of mitigating social conflict over extractive resource development.

Acknowledgements

As with any major research endeavour, many others played an important part in assisting me to complete this project. Firstly, thank you to the individuals who shared their experiences of land use conflict with me in rural communities throughout New South Wales in Australia, and Pennsylvania in the United States. Your generosity in welcoming me into your homes and workplaces to hear your stories was truly appreciated. I would particularly like to thank Kristin Saacke-Blunk as well as Peg and Bill Shuffstall for their kind hospitality and assistance while I was conducting interviews in Pennsylvania, and Holly Playford for providing logistical support and good humour on the road. For the Namoi catchment case study, I am grateful to Jacqueline Williams for her advice and assistance with interviews.

I acknowledge the support of the Australian Research Council's Discovery Early Career Researcher Award funding scheme (DE120100694), which provided the financial means for me to conduct this research project. I am also thankful to the University of New England for providing a supportive environment in which to conduct this research, particularly my supervisor at the Australian Centre for Agriculture and Law, Professor Paul Martin, whose ongoing encouragement and mentoring has enabled this research project to come to fruition.

I am grateful to several other scholars for their helpful conversations about environmental justice and land use conflict throughout the course of the research project, in particular Ted Alter, Tanya Howard, Anna Grear, Melanie Murcott, Tiina Paloniitty and Cameron Holley. In addition, I owe many thanks to my friends and colleagues from the University of New England, especially Michael Stuckey, Aileen Kennedy, Mark Perry, Lillian Corbin, Ying Chen, Eric Ghosh, Karen Lee and Mark Lunney, who offered good fellowship, sound advice and words of encouragement throughout this research project.

Lisa de Kleyn and Matthew Ryan, apart from co-authoring a chapter with me, provided helpful feedback on draft chapters and much thought-provoking discussion on the concept of environmental justice, for which I am grateful. Thanks are also due to Amy Cosby, Kylie Lingard and Michael Coleman, who also provided invaluable assistance with research for the case studies and proof-reading, and CartoGIS from the Australian National University, for their production of the case study maps. Towards the end of the writing process, Miriam Verbeek's clear thinking and editorial expertise was instrumental in renewing my focus and moving the book toward completion.

The wonderful staff at Routledge were central in bringing this book to press, and I would like to thank in particular Tim Hardwick (Senior Commissioning Editor), Ashley Wright (Senior Editorial Assistant), Amy Johnston (Editorial Assistant) and Elizabeth Spicer (Production Editor) for their patience and assistance throughout the writing and editorial process. Liz Dawn's copyediting work is also gratefully acknowledged.

Researching and writing a book involves many hours away from home, and I am grateful to the numerous friends, family and school teachers who helped out during my absences, especially while I was overseas. Specific thanks are due to Kate Wilkinson and Jo Stuckey, who assisted on several occasions.

Finally, I am forever indebted to my parents, Chris and Kathy Williamson, for their advice and support; and to my husband Patrick, whose unwavering love, encouragement and tolerance enabled me to complete this project. I dedicate this book to our two daughters, Sarah and Olivia, for providing regular (and occasionally welcome) distractions from research and writing, and for their energy and curiosity about the world that is a source of daily inspiration.

Acronyms and abbreviations

ABS   Australian Bureau of Statistics
APPEA   Australian Petroleum Production and Exploration Association
BMPA   Bulga Milbrodale Progress Association
BSAL   Biophysical Strategic Agricultural Land
CCAG   Caroona Coal Action Group
CCC   Community Consultative Committee
CICs   Critical Industry Clusters
CMAL   Coal Mines Australia Limited
CO_2 carbon dioxide
CSG   coal seam gas
DEP   Department of Environmental Protection
DGR   Director-General's Requirements
EDO   Environmental Defender's Office
EECs   endangered ecological communities
EIS   environmental impact statement
EPA   Environment Protection Agency
ESD   ecologically sustainable development
ESG   Eastern Star Gas
FOI   freedom of information
ICAC   Independent Commission Against Corruption
IESC   Independent Expert Scientific Committee on Coal Seam Gas and Large Coal Mining Development
LGA   local government area
MOU   Memorandum of Understanding
NGO   non-governmental organisation
NIMBY not in my backyard
NSW   New South Wales
PA   Pennsylvania
PAC   Planning Assessment Commission
PEL   Petroleum Exploration Licence
PUC   Public Utility Commission
REF   Review of Environmental Factors
SAL   Strategic Agricultural Land

SEARs   Secretary's Environmental Assessment Requirements
SEPPs   State Environmental Planning Policies
SIA   social impact assessment
SRLUP   Strategic Regional Land Use Policy
SSD   State Significant Development
UK   United Kingdom
US   United States

1 Land use conflict and the role of justice

Bruce Babbitt, former US Secretary of the Interior, remarked over 20 years ago that the 'next generation of environmental challenges will be more intractable, more difficult problems that fundamentally relate to how we live on the land and on the planet' (1993: 514). Current conflicts over land use involving the extraction of gas and mineral resources arguably epitomise such challenges. With the International Energy Agency (2014) estimating a 37 per cent increase in global energy consumption by 2040, resource producing countries have become concerned with how to satisfy the energy needs of a burgeoning world population, as well as guarantee their own future energy security. Despite the fact that carbon emissions from the burning of fossil fuels comprise a significant proportion of greenhouse gas pollution – and notwithstanding the considerable growth of the renewable energy sector – some nations want to boost their production of fossil fuels, particularly coal and gas resources, to meet these energy requirements. Increasing production, however, means competing for the right to extract energy resources from the land; many other land users, including agricultural and residential users, vehemently oppose proposals that impinge upon their land use practices.

Conflict over the extraction of gas and mineral resources is, of course, not a novel issue. Many studies, such as those by Hilson (2012) and Coni-Zimmer et al. (2016), have investigated conflicts in resource abundant developing countries in the global South. They found that concerns over environmental impacts; access to land and water; threats to culture, livelihood and human rights have triggered disagreements that manifest across the full spectrum of disputes, from minor protests to violent armed confrontation and warfare. Conflicts are especially prevalent in nations where weak regulatory structures and cultural, political and economic power imbalances have combined to enable unfettered development (Kemp, Owen, Gotzmann and Bond 2011). However, fewer studies have investigated land use conflict in developed countries, such as Australia and the US, even though such conflict is becoming increasingly prevalent. The rapid expansion of existing operations in these countries, combined with the search for new resources (particularly unconventional natural gas), have brought previously diffuse resource users into much closer contact with one another (Martinez-Alier 2012), resulting in new disputes. It is these conflicts that are the focus of this book.

Several factors have helped to pave the way for a resurgence of fossil fuel development in developed countries. First, new discoveries and advances in technologies, such as hydraulic fracturing and horizontal drilling, which assist in the removal of unconventional natural gas (including coal seam gas and shale bed methane), have made extractive development a commercially feasible option in locations that were previously unviable. Second, the hierarchy of property interests, which tend to privilege subsurface mineral rights over those of surface landholders, has facilitated access to both private and public land for the purposes of exploration and extraction (Duus 2013; Colvin, Witt and Lacey 2015; Van Wagner 2016a, 2016b). Third, large-scale extractive ventures have provided consumers with price competitive fossil fuel options and have also offered investment and other benefits to economically depressed communities – a boon in the wake of the global financial crisis. As Duus (2013) notes, it is frequently rural communities which are the sites of extractive development, and these areas are typically more vulnerable to fiscal pressures and population migration. Consequently, some rural residents have welcomed extractive developments, anticipating a raft of economic and social opportunities to mitigate against community decline.

The opportunities, however, are rarely realised in a socially or environmentally neutral manner (White 2013: 50). Extractive activities can create significant disruption to communities, local industries and the environment, leading many to oppose such activities. The environmental costs of extractive development are well documented, and include impacts on the quality of local air and water resources, biodiversity and climate change (White 2013). Those opposed to the development of extractive operations also raise concerns about the negative human health outcomes arising from close proximity to such operations. More and more, though, citizens have focused on the social and cultural impacts that can arise from extractive development, including:

- Threats to agricultural enterprises and farmer identity. The particular hydrogeological conditions of fossil fuel deposits often also foster the necessary conditions for high-quality agricultural land (Van Wagner 2016a). In many cases, the land targeted for mining development has been occupied for some time by rural communities engaged in agricultural activities. The fixed nature of mineral deposits means that extraction must take place where the resources are found and, as extractive activities expand, farm businesses and farmer identity come under threat.
- Dislocation of Indigenous populations who have a unique cultural and spiritual relationship with the land (Godden, Langton, Mazel and Tehan 2008).
- Increased pressure upon the physical infrastructure and social fabric of rural and remote towns resulting from fly in/fly out labour sources and associated population fluctuations (Carrington and Pereira 2011).
- Economic fallout in local areas linked to 'boom–bust' development cycles often associated with extractive activities (Brasier et al. 2011).

Government responses to land use conflict in developed countries have ranged from outright bans and moratoria on extractive development (particularly in the case of natural gas drilling involving hydraulic fracturing), through to reform of existing legal and institutional frameworks which seek to better 'balance' competing interests by promoting coexistence between different land users (Owens 2012; Duus 2013; Sherval and Hardiman 2014; Van Wagner 2016a). Yet, despite policy and regulatory intervention, many conflicts over extractive development have remained insoluble. Many disputes persist over a considerable period of time, involving multiple stakeholders and eluding resolution. Some have involved confrontation and litigation, and many have had significant spillover effects on the communities in which they are situated, taking an economic and often emotional toll on those involved. The much-publicised case of Australian cotton and grain farmer George Bender (see e.g. Whiting 2016), who took his own life as a result of lengthy battles against several energy companies that wished to pursue coal seam gas exploration on his rural Queensland property, illustrates one extreme and tragic consequence of such conflict.

The types of disputes arising from tensions between economic and energy security needs on the one hand, and local environmental, human health, and social and cultural well-being on the other, are not unique to extractive developments. Natural resource allocation determinations made by governments across the spectrum of environmental decision making, including water use, native vegetation conservation and renewable energy production, are often characterised by a clash of goals and values about what interests should be supported (Martin and Kennedy 2016). Wenz believes that these clashes have, at their heart, notions of justice. He observes:

> Many of these disputes are fostered by differing conceptions of justice. Because people have different ideas about justice, a social arrangement or environmental policy that one person considers just will be considered unjust by another.
>
> (Wenz 1988: 2)

Duus further notes that it is extraordinarily difficult to 'balance things like a stable sense of place and community, and intact and unthreatened water resources, with global energy demand, state revenue, employment, and corporate profit' (2013: 103). It may be true that conflict prevails because these divergent interests and values are 'simply incommensurate' (Duus 2013: 103; Coni-Zimmer et al. 2016); however conflict may also persist because opposing views on what makes a right or fair determination are either inadequately addressed, misunderstood or not even acknowledged in decision making processes.

The role of injustice as a driver of conflict

Decision makers often cite market efficiencies to justify decisions that impact social welfare and the environment. Such justifications, however, do not adequately address justice arguments (Syme and Nancarrow 2012: 109). Judgements about whether decisions are just or fair involve both 'rational and visceral components' (Syme and Nancarrow 2012: 108) which have a real impact upon behavioural responses to those decisions (Clayton, Koehn and Grover 2013). A number of studies show that decision making processes which are perceived to fairly consider conflicting perspectives are central to gaining acceptance of decisions and minimising disputation (often notwithstanding the substantive outcomes of decisions), while processes that are considered unfair or inequitable will cause decisions to be rejected, leading to policy failure, transaction costs and, ultimately, intractable social conflict (Martin and Kennedy 2016; Gross 2014, 2008; Clayton et al. 2013; Syme and Nancarrow 2012, 2008; Syme, Nancarrow and McCreddin 1999). Often, institutional arrangements for decision making processes concerning the development of extractive operations fail to address the:

- distribution of costs and benefits;
- availability of processes to have input into development decisions;
- adequacy of grievance mechanisms and opportunities to challenge decisions;
- constraints on landholders to veto development due to the dominance of subsurface interests;
- constraints of landholders and other stakeholders to access information;
- superior bargaining power of developers;
- privileging of technical over local knowledge;
- dialogue and conduct of government and developers;
- potential risks to other natural resources; and
- local aspirations for landscapes and communities and 'place attachment'.

To take one example: the importance of place attachment. Landscapes hold diverse, tangible values for local people and communities beyond their economic livelihood; they are a part of people's social and cultural identity, and are intrinsically connected to their physical and emotional well-being (Van Wagner 2016a, 2016b; McManus and Connor 2013; Sherval and Graham 2013; Graham 2010). Yet this social and emotional connection that people have to place is often overlooked in land use decision making processes concerning extractive development (Van Wagner 2016a, 2016b; Sherval and Hardiman 2014; Paragreen and Woodley 2013; Devine-Wright 2012, 2011a; Albrecht 2005; Duus 2013). Resource extraction is transformative in that it changes the physical landscape, in some cases to such an extent that 'a particular place is lost and another is created' (Van Wagner 2016a: 311). Such changes in physical land use can have a fundamental impact upon people's relationship with place but there is no requirement, in many jurisdictions, for a separate independent social impact assessment of place values in development approval processes (Preston 2015a). Processes focused on the economic impacts of extractive development, such as quantitative cost benefit analyses which ascribe a 'market value' to externalities, struggle to measure the true cost of social and environmental impacts, or obscure them entirely. As Sherval and Graham point out, 'land is not only profitable' (2013: 178), but decision making processes tend to constrain opportunities to raise place-based values and concerns, marginalising a tangible concern for communities.

Environmental decision makers are generally disconnected from justice; they 'do not talk the language of fairness or justice', nor do they 'apply fairness and justice principles routinely' (Gross 2014: 1). Syme and Nancarrow (2012) contend that this is in part due to a lack of decision making tools which adequately address justice concerns. It is, however, also due to a lack of coherence in explaining justice concerns in the first place. As Gross notes, the language in most environmental decision making processes 'tends to be technical and based on the specifics of the proposal, with some localised considerations of social or environmental impact' (2014: 8). Perceptions of fairness and justice are rarely given detailed attention. Decisions in favour of development are rationalised as being in the broader public interest, while local opposition may be dismissed as 'NIMBYism' ('not in my backyard'), driven by self-interest with no regard for the common good (Gross 2014; Devine-Wright 2011a, 2011b, 2012). The reductionist NIMBY label, however, overlooks divergent conceptions of what is just and fair, and buries the complexity which underlies opposition to extractive development. Motivations of self-interest are undoubtedly ever-present, but they are only 'one element in a number of issues about landscape and place disruption' (Gross 2014: 20). Perceptions of justice may be concerned with much more than just the fairness of outcomes, they can also relate to the fairness of decision making processes, how various stakeholders are treated within those processes, and how both processes and decisions affect the functioning of individuals and communities and their way of life (Clayton, Koehn and Grover 2013; Schlosberg 2007). Thus, issues of individual and group identity, status and inclusion come into play when stakeholders assess the processes and outcomes of environmental decision making (Clayton and Opotow 2003).

Governance arrangements – including laws, policies and the institutional structures charged with their implementation – that either overlook or marginalise the variety of justice concerns within situations of conflict are 'unlikely to be socially optimal or robust over time' (Martin and Kennedy 2016: 109). In order for responses to land use conflict over extractive development to be 'meaningful and lasting', they must 'move beyond technical and procedural advances, and engage with deeper questions' (Duus 2013: 97). This means seeking a greater understanding of the justice issues inherent in such conflict. Bringing the concept of justice to the fore – how it is perceived, and why it matters – can provide a path forward to resolving conflict by developing critical insights into what drives disputes, and how social acceptance of decisions can be established (Gross 2014; Walker 2012; Syme and Nancarrow 2012; Van Wagner 2016a, 2016b). But this is not a simple task. Perspectives on justice are 'frequently intertwined' and may be 'difficult to understand or tease apart' (Gross 2014: 8–10). Indeed, justice issues may not even be well understood or even specifically acknowledged by those affected, despite a strong sense of *injustice* (Coni-Zimmer et al. 2016; Gross 2014). To better understand how justice issues influence land use conflict, scholars have advanced the concept of environmental justice.

Initially, the concept of environmental justice was used as a way of framing complaints against the disproportionate siting of toxic waste facilities and other environmental harms near racial minorities and other disadvantaged groups in the US (Bullard 1990; Pulido 1996). However, it has since grown to incorporate a wide spectrum of environmental issues, and is now used all over the world as 'an important way of bringing attention to previously neglected or overlooked patterns of inequality which can matter deeply to people's health, well-being and quality of life' (Walker 2012: 1). In Chapter 2, we will discuss in detail the various articulations of the concept. In brief, environmental justice refers to a range of justice issues, including the ways in which environmental harms and risks are distributed throughout society, opportunities to participate in environmental decision making, the recognition and respect of those impacted by environmental harm and, ultimately, the functioning and flourishing of individuals and communities (Schlosberg 2007). In other words, environmental justice includes the array of justice concerns which typically occur and overlap in cases of land use conflict.

Researchers on extractive industries and land use conflict in developing countries have, for some time, drawn upon environmental justice as a means to explore and explain claims for justice (Urkidi and Walter 2011; Tschakert 2009; Martinez-Alier 2001). Increasingly, researchers on land use conflict and extractive activities in developed countries have also used the concept of environmental justice as a means to understand disputes (White 2013; Perry 2013, 2012). However, while there are some exceptions (notably Van Wagner 2016a, 2016b, and Fry, Briggle and Kincaid 2015), much of the empirical work on the concept has focused on singular aspects of environmental justice, particularly on tracing the substantive distribution of environmental harms and risks (Clough and Bell 2016; Ogneva-Himmelberger and Huang 2015). My aim in this study is to understand the full range of justice concerns which emerge in cases of intractable land use conflict over extractive development in rural communities within developed countries; drawing on the theoretical underpinnings of environmental justice (Chapter 2) as well as real-world studies of conflict (chapters 3 to 6) to provide an insight into why these conflicts develop, and how they might be better managed.

The environmental justice framework

A central argument in this book is that focusing on claims for environmental justice can provide insight into 'how people think, reason and act in relation to environmental concerns' (Walker 2012: 1) and, ultimately, what drives conflict. But this is not a straightforward task, because land use and resource allocation disputes are contentious and 'wicked' social problems, comprised of many 'interests, perspectives, experiences, knowledge and contexts' (Gross 2014: 40). As already explained, claims of injustice in land use conflicts typically involve assertions about the unequal impacts that specific land uses have on particular individuals or groups, but they may also advance broader views about the 'right' ways to allocate land use within society. Claims are also framed by various parties, who will offer distinct interpretations of how things are and how they should be, and different types of evidence to support their claims (Walker 2012).

In order to understand the diverse claims for justice I encountered in my research, I found it helpful to utilise Walker's (2012) simple yet comprehensive framework for making sense of claims for environmental justice. He proposes that environmental justice claim-making is made up of three elements: justice, evidence and process (see Figure 1.1).

Justice refers to norms or 'how things ought to be'; evidence is descriptive, detailing 'how things are', what is considered unequal between different groups and how it is experienced; and process seeks to explain 'why things are the way they are' through identifying the causes of injustice. The combination of evidence of inequality with claims of what is just forms the core of environmental justice claim-making; '[o]ne without the other is undoubtedly less complete and typically less effective in providing a case for action, change or redress' (Walker 2012: 53). Accordingly, Walker places these elements on the main axis at the top of the illustration in Figure 1.1, indicating that at the centre of a claim for environmental justice must be evidence of inequality as well as reasoning as to why that inequality is wrong. Linked to the main axis is the third element of environmental justice claim-making: process. The process element seeks to explain 'why things are the way they are' through identifying the causes of injustice. Causes of injustice may be context-specific, for example focused on localised decision making practices, or structural and linked to more systemic issues, such as the distribution of power in society. Walker believes that understanding each element is necessary to reveal the 'explicit and implicit meanings, values and ideologies' (2012: 39) which underpin conflict.

Walker notes that much of the academic literature on environmental justice tends to focus either on normative justice theories and philosophies (justice), or on empirically detailing the distribution of inequalities (evidence) (2012: 6). But evidence of difference or inequality is not in itself sufficient to demonstrate injustice; because justice is a contested notion, what is considered in one place to be unequal will not necessarily be considered so elsewhere. Walker argues that

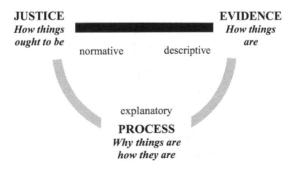

Figure 1.1 Walker's three elements of environmental justice claim-making (from Walker 2012: 40)

claims of inequality should be accompanied by a judgement as to the nature of the difference, why it matters and how it might be remedied. His framework encourages us to bring together the 'is–ought' – the description of the inequality that *is* experienced as well as an evaluation of what the situation *ought* to be, measured against normative qualities of justice. This conception is particularly useful when examining claims for justice concerning extractive development in more developed countries where, often, those seeking justice (e.g. farmers and rural residents) do not necessarily fit within the typical categories of environmental justice claimants (e.g. racial minorities, or the poor). As Walker ponders of a case study involving waste facility siting in Scotland: 'Could a community surrounded by landfill waste sites . . . still be a case of "environmental injustice" even though it was predominantly white and lower-middle class in social make-up?'. Answering such a question requires a more nuanced and reasoned understanding of why difference is a concern, and how it is being produced. The 'is–ought' distinction – by being 'descriptive with inequality and normative with justice' (Walker 2012: 13) – helps to not only reveal inequality but also *why* it is wrong and *why* it occurs.

Walker's framework does not prescribe a particular sequence that environmental justice claim-making must follow. Indeed, his diagram (Figure 1.1) is absent directional arrows to indicate (for example) that evidence comes before a consideration of normative justice concepts, to be followed by a determination of why injustice exists. There is, in reality, a 'cycling and circling' (2012: 41) between each of the elements as claim-making progresses. In the subsections below, I discuss the three elements of claim-making in further detail. The discussion of theories of environmental justice, in the first subsection, introduces some of the key concepts which are further detailed in Chapter 2. In the second subsection, I focus upon the element of evidence and, specifically, upon the methods employed in this research to gather and analyse data. In the third subsection, I introduce the process element, which is also discussed further in Chapter 2. Throughout chapters 3 to 6, I link empirical evidence from the case studies to the literature on justice and process to offer explanations as to why conflicts over extractive development have arisen, and how they might be better managed.

Theories of justice: 'how things ought to be'

The concept of environmental justice draws upon theories of justice to articulate a normative position against which claims can be measured. Theories of just distribution – or who gets what – have tended to dominate definitions of environmental justice. This is because the initial focus within the environmental justice movement was on the siting of environmental harms, highlighting instances of maldistribution where certain vulnerable groups (including communities of colour, Indigenous groups and the poor) were disproportionately exposed to environmental risks and harms (such as toxic waste dumps and other industrial pollution) (Bullard 1990; Schlosberg 2007; Walker 2012; Bell 2004).

As the environmental justice movement grew and moved beyond the US to other countries, the scope of 'vulnerability' extended to include other contextually relevant factors, such as socio-political disadvantage (Walker 2009, 2012). The notion of what constitutes 'justice' also evolved, with increased attention to the processes that determine distributive outcomes, and how these outcomes impact upon the well-being and functioning of individuals and communities (Schlosberg 2007).

Procedural justice looks beyond substantive outcomes to the processes and 'institutional contexts' (Shrader-Frechette 2002: 27) within which distributive decisions are made. It is concerned with access to (and the nature of) participation in decision making processes. In the environmental realm, this can include decisions over land use and the siting of facilities that will have an impact on natural resources and the environment. Researchers have developed various criteria for recognising procedural justice in environmental decision making (see e.g. Hunold and Young 1998; Smith and McDonough 2001; Bryner 2002; Maguire and Lind 2003; Arcioni and Mitchell 2005; Millner 2011; Preston 2015b), including:

- the adequacy of notice of upcoming decisions;
- the availability of information about decision making, including the decision making processes, as well as its effectiveness (e.g. whether it is timely, affordable, easily accessible);
- consultation and inclusion in decision making processes;
- access to independent and impartial review bodies (such as courts) to challenge decision making processes; and
- broad interpretation of standing rules to access courts and tribunals to challenge environmental decisions.

A lack of procedural justice may not only lead to distributive injustice but is also a source of injustice in itself (Walker 2012). Hunold and Young argue that such procedural conditions are necessary because it is unjust to impose environmental risks upon citizens without enabling them to participate in decision making processes (1998: 83). They note further that fair processes result in more thorough and considered determinations and can even ensure that distributively fair outcomes are achieved (1998: 87). The importance of ensuring procedural justice in environmental decision making is increasingly acknowledged in international instruments, as well as in many national laws and policies.

Procedural requirements provide 'ideal' conditions against which claims for procedural justice may be measured. Yet, the existence of procedural conditions alone may not be sufficient to ensure justice because 'unequal power relations and the authority to make decisions are not so readily put aside' (Walker 2012: 49). In other words, despite efforts to overcome procedural barriers, effective participation may be hindered by the ways in which decision making processes are conducted, such as whether participants are treated fairly and respectfully (Young 1990; Fraser 1998, 2000). Schlosberg makes the case that 'recognition' must be included in our understanding of environmental justice because a lack of recognition – in the form of 'insults, degradation, and devaluation at both the individual and cultural level' – not only constrains participation but also leads to distributive injustice (2007: 14).

Recognition has been theorised as an element of justice in several ways: first, from an individual psychological perspective of self-worth, where misrecognition and disrespect are seen to cause humiliation and harm to self-identity (Honneth 1995; Bies 2001); and, second, from a structural status-based perspective, which sees injustice in cultural domination, cultural invisibility ('non-recognition') and cultural disrespect (through stereotypical depictions, such as dismissing groups opposed to development as NIMBYs) (Fraser 1998). Schlosberg (2007) suggests that both the psychological and structural dimensions of recognition may be simultaneously evident in claims for environmental justice and advocates for a broad understanding of recognition – alongside procedural fairness and distribution – in his definition of environmental justice.

Increasingly, claims for environmental justice have centred more generally upon the well-being of individuals and communities, and the sense of injustice felt when their capacity to function is constrained. While theories of distribution, procedure and recognition offer some insight into specific aspects of justice, Schlosberg argues that a fourth theoretical category of justice – capabilities – can provide an integrative framework to bring together the full range of concerns. As he notes, it may be convenient to discuss the various theoretical aspects of justice in isolation but the experience of injustice overlaps and is felt in various ways at the same time:

> While equity is discussed, recognition is demanded, and participation in policymaking is sought . . . campaigns in the environmental justice community are more generally about re-establishing the capabilities necessary for a healthy, functioning community.
>
> (Schlosberg 2007: 72)

The capabilities approach looks to whether individuals and groups possess the freedom to do and be what they want. It measures whether social arrangements foster the necessary capabilities to enable desired functionings, such as having a healthy environment, or being well nourished. The capability to achieve valued functionings is 'where the space for determining justice is located' (Walker 2012: 52); injustice is found where a capability (such as political freedom) is either restricted or denied (Schlosberg 2007). The metric of justice is what individuals and groups are able to do, and whether social arrangements expand or contract that ability. While the capabilities approach is more prevalent in studies of economic development, it has been recently used to explore claims of environmental injustice (Groves 2015; Griewald and Rauschmayer 2014; Tschakert 2009).

A distinct advantage of bringing capabilities into our understanding of environmental justice is that it recognises the interdependency of human and natural systems, and the potential for injustice in the non-human realm. While environmental justice scholarship focuses on environmental hazards and impacts, much of it is typically connected to (and comprehended in terms of) injustice to human populations (Jessup 2015; Neimanis, Castleden and Rainham 2012). This anthropocentric approach is distinct from more 'eco-centric' conceptualisations of justice to nature and ecological integrity (Jessup 2015; see also Westra 1994; Plumwood 2001). A capabilities approach to environmental justice recognises that ecological functioning is critical to human functioning. Regardless of whether environmental impacts cause injustice to humans, impaired ecological capability will struggle to support human life (Schlosberg 2007; Neimanis et al. 2012). Making this connection acknowledges that the 'human dimensions of environmental justice are embedded in just relations with the "more-than-human" world' (Van Wagner 2016a: 314). The capabilities approach expands our understanding of environmental justice beyond human experiences of injustice, and can assist in bringing into focus the 'overlapping and nested relationships within the ecology of specific landscapes' (Van Wagner 2016b: 40).

Evidence: 'how things are'

Theories of justice have a great deal to tell us about inequality but they are only part of the picture. We must consider notions of justice against everyday experiences and perceptions of what is 'just'. This speaks to the second element of Walker's (2012) claim-making framework, evidence; that is, what is considered unequal and how it is experienced. The combination of evidence with normative theories of justice comprises the 'core' of environmental justice claim-making and both are required in order to advance a case for change or rectification (Walker 2012: 53).

Throughout this subsection, I detail how evidence was gathered for this research, including the methods employed to identify and analyse evidence. As Walker (2012) notes, the forms of evidence utilised will vary depending on the nature of the claim advanced; evidence of difference may be quantitative (such as statistics which detail patterns of distribution of environmental harms and risks), or qualitative (such as observation and interview accounts). The critical point is that evidence can come from various sources of data, and data produced by lay people, such as narratives which detail their values or how they have been treated and how they feel, is equally as valid as technical and scientific data from experts (Walker 2012).

Given that my aim in this study is to explore how governance mechanisms either alleviate or exacerbate cases of conflict over extractive development, I looked to the sites of conflict that first drew my attention to this topic of conflict and justice. Extractive operations are interwoven within the places they are situated; they are, as Ey and Sherval note 'complex socio-cultural terrains':

> The performance of resource extraction is inseparable from the situated socio-cultural contexts in which it transpires . . . extractive terrains must be perceived in terms of their capacities to both shape and be shaped by their

socio-cultural contexts, and in terms of their ability to form a critical part of individual and collective identities across a range of scales and temporalities.

(Ey and Sherval 2016: 178)

Within these socio-cultural contexts, the law emerges as a pivotal factor in shaping the ways in which extractive industries operate. Laws, regulations, policies, administrative decisions and court cases are 'crucial sites of contestation over human–environment relations' (Andrews and McCarthy 2014: 9); they have 'specific and material consequences' for both people and places by determining the boundaries of disputes, and their consequent outcomes (Van Wagner 2016b: 37; Martin and Kennedy 2016). However, while the law may 'bear all the marks of a rhetoric of impersonality and of neutrality' (Bourdieu 1987: 819) it is also not independent of its social context; it is constructed, applied and experienced within particular *places*. The law creates and gives meaning to places, and places, in turn, create and give meaning to the law; they *co-produce* one another. Accordingly, efforts to understand the role of the law in shaping and constraining conflict over extractive development must also acknowledge the political, economic and socio-cultural contexts within which legal instruments are embedded. Thus, relevant questions include: How are laws created and interpreted? Who establishes the construction of specific legal meanings? How does the law interact with other non-law variables?

Scholars have become increasingly aware of the linkage between law and space – its 'situatedness' and 'terrain' – and have sought to explore the causes and effects of law by foregrounding its spatial context (Philippopoulos-Mihalopoulos 2011: 187; 2014: 15; Graham 2010; Delaney 2010). Legal geographers, in particular, have paid attention to the dynamic relationship between law and space, seeing the law as a social discourse that both generates and responds to places (Turton 2015; Bartel et al. 2013; Benson 2014; Blomley, Delaney and Ford 2001). By making the spaces in which the law operates and the relationships it creates an 'active factor' (Philippopoulos-Mihalopoulos 2010: 206), rather than simply 'an inert backdrop against which legal actors play out their disputes' (Aoki 2000: 937), we can see how the law is produced and deployed to channel conflict. By foregrounding the lived experience of law *and* its contextual space – how they produce and give meaning to one another, how they both construct and are influenced by relationships of power – we are able to glean a richer understanding of justice, injustice and patterns of conflict.

Capturing the 'constellation of laws, actors, institutions and dynamics' (Apple 2014: 243) relevant to conflict over extractive development required me to adopt a research approach that detailed how legal, social and spatial relations are 'weaved and re-weaved . . . from the site up' (Bennett and Layard 2015: 410). Accordingly, I chose an in-depth qualitative case study approach. The case study method is particularly suited to answering 'how' and 'why' questions, providing the researcher with the opportunity to develop a rich understanding of dynamic social relationships and contexts (Yin 2009; Stake 1995). Although, as Nagle points out, 'no set of case studies can yield definitive proof about the operation of the law' (2010: 8), in-depth case studies of places and how the assemblage of legal instruments, institutions and actors function within them can offer a detailed account of the complexities encountered in conflict situations. They can reveal experiences and expectations of justice in cases of land use conflict that 'encounter the law' (Jessup 2013: 83), highlighting legal norms as well as people's perceptions and motivations for behaviour, relationships of power and the 'paths to influence' (Jarrell, Ozymy and McGurrin 2013: 188). Overall, the in-depth case study approach reveals the 'complexities and contradictions of real life' – the 'rich ambiguity of politics' along with the 'minutiae' that is 'pregnant with paradigms, metaphors, and general significance' (Flyvbjerg 2006: 237–238).

For my research, I chose a multiple case study design, which enabled me to compare issues of fairness and justice in land use conflict across distinct settings, different jurisdictions (and hence legal frameworks), different community contexts and different resource extraction issues (open cut coal, coal seam gas, shale gas). The cases I selected for this study have strategic significance in relation to the problem of land use conflict concerning the development of extractive industries. In each case, there had been a period of prolonged public opposition to extractive development, along with specific legal and policy interventions that were implemented to manage the approval of extractive activities. In each case, notions of environmental justice had been variously deployed to articulate community perceptions of unfairness and injustice.

I collected data for each case study from a variety of sources: semi-structured interviews with key informants, legislation, case transcripts, policy documents, records of public meetings, development applications, submissions, media releases and news articles.

The interviews formed a significant portion of the data collected because they provided the best opportunity to probe into perceptions of fairness and justice in cases of land use conflict (Gross 2008). Interviewing is arguably 'the central resource through which contemporary social science (and society) engages with issues that concern it' (Rapley 2001: 303–304). I used a predetermined set of questions, which drew upon the theories of environmental justice, to help structure the interviews. Interviews began with questions about the background and setting of the case study, including the role of the interviewee and their knowledge of the relevant law and policy context, before moving on to the interviewee's perception of the conflict situation and concepts of justice and fairness. Interviewees were asked for their suggestions about what might have been done differently, particularly where an injustice was perceived. It should be noted that I did not strictly follow the questionnaire, opting for a 'conversational and informal' tone (Longhurst 2010: 105), and exercising flexibility in altering the line of questioning or to seek further elaboration (May 2011). This style of interviewing enabled me to establish rapport with interviewees as well as maintain sensitivity, which was critical given the subject matter; questions concerning fairness and justice 'tap into people's innermost feelings' (Gross 2008: 46).

In total, I conducted 46 interviews with stakeholders across the three case study sites. In most cases, I conducted these interviews face to face. On some occasions, where it was not convenient to conduct an interview in-person, I conducted the interview over the telephone. Each interview typically lasted between one-and-a-half and two hours. I selected interview participants on the basis of their involvement in the case study situation, including community members, industry representatives, representatives of non-government organisations (NGOs) (including public interest environmental lawyers), as well as staff from government offices. I used a process of accumulative chain (snowball) sampling (Biernacki and Waldorf 1981; Noy 2008) to locate potential participants through other informants. Interviews were recorded to allow me to focus fully on the discussions and interactions, though I also took field notes (Longhurst 2010; Yin 2009). These notes became a part of the data set for the research, and were a useful reflective device when reviewing transcripts of the interviews.

Documentary evidence supplemented the interview data:

> Court documents frame the legal issues, government policies and assessment reports expose the assessment scale and project priorities, environmental assessment documents detail the project and its impact, submissions outline the support and opposition to the project, and news reports capture emotional moments in the dispute that show changes in the attitude of the community and the proponents.
>
> (Jessup 2013: 74)

I used a thematic discourse analysis, which categorises data according to 'repeated patterns of meaning' (Braun and Clarke 2006: 86), to examine both the interview and documentary evidence. Such an analysis seeks to explore how language is used in particular contexts to construct social realities; that is, 'what specific version of the world, or identity, or meaning is produced by describing something in just that way over another way; what is made available and what is excluded by describing something this way over an alternative way' (Rapley 2007: 2). Law and space are not only material but constructed discursively and performatively, so I focused on law and space as social action (Willig 2003: 165). As a 'relational' form of research, the emphasis of discourse analysis is not on 'entities or individuals', but on social relations more broadly (Fairclough 2013: 3).

I adopted Braun and Clarke's (2006) six steps for conducting the thematic analysis, beginning with an exploration of interview transcripts and documentary evidence to look for 'patterns of meaning and issues of potential interest in the data' (2006: 86). I developed thematic 'codes', including key words and implicit constructions relevant to the research question, organising the data into meaningful clusters. I paid particular attention to the various 'dimensions of discourse': how did the law and legal institutions construct their 'objects and subjects', and what were the consequences of their deployment in specific ways (Willig 2003: 166)? As I became increasingly familiar with the data, I identified new codes and themes, which fed back into subsequent stages of analysis. The process of thematic analysis was therefore not a linear one; the analysis was 'a recursive process' and involved 'a constant moving back and forward between the entire data set, the coded extracts of data . . . and the analysis of the data' (Braun and Clarke 2006: 86).

Overall, I was mindful that evidence is produced and understood within social processes, and may be, as Shrader-Frechette asserts, 'incomplete and saddled with implicit interpretations' (2002: 194). Evidence 'is not something given, a set of truths waiting simply to be revealed' (Walker 2012: 41); it is created by a number of sources, shaped by power relations between actors, and given meaning within a social context. Thus, I attempted to maintain an essential 'healthy scepticism' about evidence and its role in claim-making (Walker 2012: 41).

Process: 'why things are how they are'

The final element of Walker's framework, process, seeks to connect theory with practice by explaining why inequalities exist, and why they are unjust. There is some overlap here with the theoretical justice concepts discussed earlier, because distribution, participation and recognition failures are not only forms of injustice, they can also be linked to other forms of injustice. For example, a lack of participatory opportunities is itself an injustice, but can also explain distributional inequality; similarly, a lack of recognition is not only a distinct inequity, but can also erode meaningful participation in decision making (Walker 2012; Schlosberg 2007).

Beyond these connections, other explanatory claims have also been advanced which further detail the causes of environmental injustice. As noted above, environmental racism and discrimination was an early dominant explanation of environmental justice in the literature; however, as the scope of environmental justice expanded to other situations of difference and disadvantage, scholars adopted additional frameworks from social and political theory to explore and explain the causes of environmental injustice (Walker 2012; Brehm and Pellow 2013). From the explanatory frameworks that have been articulated in the literature, power emerges as a key theme in the generation of environmental injustice. As Walker explains, power can take a number of forms, including structural (the power of economic and political frameworks), material (the power of certain objects) and discursive (the power of language) (2012: 72). When combined, these 'power geometries' decide who controls and who is excluded from environmental decision making (Heynen, Kaika and Swyngedouw 2006; Walker 2012). However, power can be hidden and difficult to detect, and may be mobilised in subtle ways to shape how conflict is viewed and resolved (Gaventa 1980; Lukes 2005); for example, through the construction or prioritisation of a particular scale of assessment (e.g. regional) which excludes other scales of concern (e.g. local) (Huber and Emel 2009; Jessup 2013); or even in the ways in which certain forms of evidence and knowledge are valued or devalued (Martin, Gross-Camp, Kebede, McGuire and Munyarukaza 2014; Walker 2012). In seeking to understand matters of process, tracing 'how, where, and by whom power is used' (Andrews and McCarthy 2014: 9) is critical to unravelling how environmental injustice is constructed.

Recent empirical analyses of land use conflict have explored how the law is used by those in power to shape the social and legal spaces of disputes. In one study, Andrews and McCarthy (2014) explore the spatio-legal dynamics of shale gas development in Pennsylvania (which is also the subject of the case study presented in Chapter 6). The authors were concerned with not only the formal legal and policy arena but also the informal and 'extra-legal' dynamics that shaped the introduction and interpretation of particular laws. They found examples of power operating 'through the law', where strategic legislative interventions were used to centralise and reinforce government authority and subvert local control. Other contextual factors – historical, economic and political – combined to facilitate the rapid expansion of shale gas development.

In another case, Jessup (2013) similarly demonstrates how the law relevant to the approval of a waste facility in NSW, Australia, not only constructed but strategically prioritised certain spaces and scales in order to secure development consent. Power relations facilitated the imposition of an ideological notion of 'public interest' at the regional scale, displacing local concerns about environmental impacts, and disempowering the community.

Most recently, Van Wagner (2016a, 2016b) has demonstrated how, in the case of aggregate extraction in Ontario, Canada, existing legal structures privilege particular relationships with the land; namely, property owners'. As she notes, this 'obscures "more-than-ownership" relations with the land', including complex connections to place (2016b: 57).

In each of these case studies, the environment was not only 'an object of governance', but became a 'terrain of struggle with respect to the law' (Andrews and McCarthy 2014: 7). This book explores the same phenomena in cases of land use conflict over extractive development. It focuses on the interrelationships between power and place, and the ways in which governance mechanisms structure 'people–place relations' in order to 'mediate spatio-legal production in struggles over land use' (Van Wagner 2016b: 41).

Beyond the centrality of power to the theoretical explanations of process which have emerged in the literature, other commonalities are also apparent. First, injustice defies a simple explanation because it is the result of 'multiple interwoven processes' (Walker 2012: 74), which may result in the experience of numerous and different forms of inequality at the same time. Second, given the complexity of the processes which produce inequality, there is a need to explore their history and operation over time to fully comprehend the causes of injustice. Finally, while claims about local processes are grounded in a specific context, they are often linked to broader structural processes at regional, national and even global scales (Walker 2012: 74). This re-emphasises the need for a broad and plural theory of justice and a critical approach to the evaluation of evidence in seeking to understand claims for environmental justice.

Conclusion

Intractable conflict over the extraction of natural resources, including coal and unconventional gas, has recently emerged as a pressing policy concern in several rural communities throughout developed countries. While conflict of this nature is not new, the rapid expansion of extractive development in areas occupied by other land users – in particular, farmers and rural residents – has recently witnessed an acceleration in the scale and intensity of such disputes.

In this introductory chapter, I have highlighted that conflicts over land and resource allocation can become insoluble when efforts to manage them neglect the different perceptions of justice which are often at the heart of these disputes. It is a fact that an ever-expanding global population demands increased energy, and tough decisions must be made about land use and resource allocation. However, a deeper understanding of what is essentially at stake when resource use decisions are being made could minimise conflict. Accusations of self-interested NIMBYism, often used to characterise objection, fail to capture the extent of justice concerns. Conflicts over extractive development reveal a confluence of interests and concerns which extend beyond the substantive outcomes of disputes. These include whether communities and individuals impacted by resource extraction have meaningful opportunities to participate in decision making, how they are treated in decision making processes, and, ultimately, whether decision makers adequately account for impacts upon community well-being and their 'way of life'. Governance interventions that misjudge or overlook these concerns are destined to fail, inevitably leading to conflict.

The concept of environmental justice provides a mechanism for investigating these complex issues, helping us to comprehend why conflicts become resistant to legal and policy intervention and how they could be differently managed. Chapter 2 further explores the concept of environmental justice. Following Walker's (2012) framework for making sense of claims for environmental justice, Chapter 2 focuses specifically on theories of justice, and process explanations for the causes of injustice. This review of the literature provides a range of concepts and ideas that are useful for considering empirical evidence of inequity. These themes are subsequently drawn upon throughout the investigation of real-world cases of land use conflict over extractive development (described in Chapters 3 to 6), exploring how justice (and injustice) is perceived, and how these perceptions influenced situations of land use conflict. By placing justice as a central concern, I uncover how a sense of injustice motivated these disputes, and reveal why governance mechanisms have not been effective in overcoming land use conflict.

When we better understand the factors that drive conflict over land use, we can begin to reconsider approaches to land use decision making practices. The concept of environmental justice opens up a dialogue about alternative approaches to environmental decision making that can result in outcomes that are better informed, and more likely to be fair. In the final chapter, Chapter 7, I use lessons learnt from the case studies to propose potential directions for reform. Justice-based approaches to land use decision making are more likely to mitigate broad-scale social conflict over extractive development. In closing this book I advocate further research to explore the possibilities open to decision makers.

2 Theories of environmental justice

With Lisa de Kleyn and Matthew Ryan

Introduction

The fundamental proposition of this book is that justice is central to cases of land use conflict concerning resource extraction and development. Before continuing with evidence for this proposition (provided in Chapters 3 to 6), it is necessary to discuss how the concept of justice is conceived in this book. Justice is a multifaceted concept – one that has been theorised and analysed throughout history and across many disciplines. It has 'different meanings for different people' (Tschakert 2009: 731) and is 'situated and contextual, grounded in the circumstances of time and place' (Walker 2012: 11).

In this book, 'environmental justice' is used to bring together key justice themes relevant to resource allocation and land use as a 'broad, overarching concept encompassing all justice issues in environmental decision-making' (Ikeme 2003: 200). This chapter elaborates how justice has been understood within the environmental sphere, and how the causes of injustice have been explained. Schlosberg (2007) proposes a multivalent conception of environmental justice, incorporating the dimensions of distribution, participation, recognition and capabilities. This chapter draws particularly on Schlosberg's (2007) multivalent conception of environmental justice to elaborate on how justice has been understood within the environmental sphere, and how the causes of injustice have been explained.

Environmental 'justice'

The environmental justice movement (and associated scholarship) applies justice theories to the exploration of environmental conditions. It is predominantly concerned with the identification of *injustice* in environmental policies, institutions, social structures and outcomes. The concept is different from similar, more-visible terms, such as 'sustainability', which are often used in the rhetoric for legitimising current practices. Generally, environmental justice is a concept adopted almost exclusively by those criticising the status quo.

In the literature on environmental justice, there is general agreement that the 'environmental' component of the concept should be construed widely and not restricted to wilderness areas (Schlosberg 2007, 2013). The environment is understood as a part of people's everyday lives – where they 'live, work, and play' (Novotny 1995; Schlosberg 2007). Such a conception creates a holistic view of the human–nature relationship that is not intended to be anthropocentric, and acknowledges the inextricable link and dependency between human and physical environments. The 'justice' component of the environmental justice concept is what has attracted the most analysis and debate.

At first, the concept of justice appears to be simple, synonymous with fairness – indeed, in *The Road to Justice*, Lord Denning defines justice as 'what the right-minded members of the community – those who have the right spirit within them – believe to be fair' (Denning 1955: 4). Fairness is, in turn, a seemingly ubiquitous concept, appreciated almost innately by children. Even for children, however, it is *scarcity* that problematises fairness. It is only when there is a relative shortage of a certain good that fairness becomes an issue – only when there is one biscuit to share among two children, for example. If something – biscuits or environmental goods – is infinitely available, then parties will consume their fill, without concern for justice:

> Justice usually becomes an issue in a context in which people's wants or needs exceed the means of satisfaction . . . In these situations, people are concerned about getting their fair share, and arrangements are made, or institutions are generated, to allocate scarce things among those who want or need them.
>
> (Wenz 1988: 6)

Perceptions of just distribution are crucial for maintaining social cohesion. As Wenz notes, 'social solidarity and the maintenance of order in a relatively free society require people to consider their sacrifices to be justified in relation to the sacrifices of others' (1988: 21). Distribution thus dominates most discussions of justice; it is 'an idea with a very long pedigree', with origins in the writings of Aristotle and, later, Aquinas (Miller 1999: 2). In more contemporary theory, Rawls's influential *A Theory of Justice*, defines justice as 'a standard whereby the distributive aspects of the basic structure of society are to be assessed' (1971: 9–10). Efforts to define justice within the distributive paradigm have thus largely focused upon how to structure the allocation of substantive benefits and burdens within society; such a focus also occurs within the concept of environmental justice.

Distributive *environmental* justice is concerned with the equitable sharing of environmental benefits and burdens across the population. A significant focus of much of the early environmental justice research was upon identifying patterns of uneven distribution among vulnerable groups, including Indigenous communities, communities of colour and the poor, who have been repeatedly found to be disproportionately impacted by environmental risks and harms (Schlosberg 2007; Figueroa 2003). Over time, however, the focus on proximity to harm has been broadened to other categories of disadvantage, most notably political inequality (Walker 2009, 2012). Environmental justice is, therefore, no longer exclusively concerned with tracing the distribution of harm within impoverished and minority neighbourhoods; it can be 'analysed within and between "communities" (however defined) where one community is more vulnerable to environmental deterioration than the other' (Jessup 2013: 77). Accordingly, environmental justice scholarship has more recently turned its attention to spaces of political power and vulnerability within and between different scales (e.g. local/regional/state/ national) (Jessup 2013; Walker 2009).

Within environmental justice scholarship and activism, claims about distributive justice are typically concerned with 'who, what and how' – who constitutes the community of justice; what environmental benefits and burdens are to be distributed; and the principles determining how distribution will take place (Bell 2004; Walker 2012):

- The recipients of environmental justice – or 'who matters' (Walker 2012: 42) when it comes to distribution – may be a readily identifiable current human population of a particular country or it may extend beyond spatial and temporal boundaries to include a broader geographical area, future generations or non-anthropocentric populations (Bell 2004).
- Ascertaining 'what' is to be distributed requires understanding the concepts of benefits and burdens relative to the identified community of justice, and deciding upon the relevant metric (Walker 2012). While 'what' is to be distributed has largely been expressed in terms of environmental harms and risks (e.g. pollution, flooding), more recent research has included access to environmental 'goods' such as quality environmental spaces, and other environmental resources including food and fuel (Bell 2004).
- Defining the principle of distribution involves deciding what constitutes the best principle for the allocation of environmental benefits and burdens. There are many perspectives from justice scholarship on appropriate principles of distribution. Over time, there has grown a general agreement that the equal distribution of environmental burdens should be complemented by guaranteed standards, shifting the focus from the equal distribution of 'bads' (such as pollution) towards rights to environmental benefits (such as clean water and air) (Bell 2004: 294). With varied conceptualisations of 'what' should be distributed (in particular, *access to quality environments*), this focus on rights has more recently extended to include a minimum standard above which individuals can exercise preferences (Bell 2004).

In practice, determining the principle of distribution poses complicated questions, and the implications that emerge will differ depending on the preferred conception of distribution. For example, utilitarianism is often used to justify siting decisions. It calls for the distribution of benefits and burdens in a manner that results in the greatest outcomes for the greatest number, and informs much environmental and natural resource law and policy (Bryner 2002: 41). But this approach, in practice, quickly devolves into conventional cost–benefit analysis, raising significant quantitative and qualitative challenges. As Preston (2015b: 26) argues, cost–benefit analyses are typically not concerned with fair distribution but with the aggregation of costs and benefits. The benefits of consuming uses of the environment (such as coal extracted or timber felled) are much easier to quantify in monetary terms, while the burdens of consumption upon ecosystem health, public health and aesthetic impacts are more difficult to calculate and often end up underestimated. In addition, just as burdens are often underestimated, the broader economic benefits of resource consumption are often severely *overestimated*. Moreover, environmental laws which facilitate approval for consuming uses also tend to place the burden of proof on those seeking to preserve the environment rather than on consuming users to demonstrate that there will be an absence of environmental harm, creating a presumption of approval and further skewing distribution (Preston 2015b). Alternative conceptions of distributive justice, such as merit-based approaches, the polluter pays, ability to pay and beneficiary pays principles, and prioritising disadvantage (Hillman 2006; Dietz and Atkinson 2005) face their own challenges. The question remains: Who should bear the cost of environmental harm and how should it be valued?

The challenge of deciding who should bear the cost is a significant factor in the case studies discussed in this book, which show that extractive activities produce intense local impacts that are disproportionately felt by those situated in close proximity to development. Principles of just distribution require that these impacts be shared equally but, unlike the siting of other environmental hazards, such as waste processing facilities which may be 'intentionally "spreadable" over space' (Walker 2012: 150), the siting of extractive activities is dictated by the 'uneven geography' of where geological deposits happen to be located (Huber and Emel 2009). Remedying unjust distribution essentially becomes a question of scale: at what level – local, regional, state, national – should the benefits and burdens of resource extraction be measured?

Concepts of distribution are central to environmental justice. However, conceiving of justice solely in distributive terms may be inadequate to capture the variety of concerns at stake. Other concerns include opportunities for public participation, respect and recognition of individuals and communities, and the capacity of individuals and communities to function fully (Schlosberg 2007; Figueroa 2003; Shrader-Frechette 2002; Cole and Foster 2001). In many cases, concerns move beyond anthropocentric ones, linking community capabilities with issues of ecological sustainability (Schlosberg 2007). Throughout the history of the environmental justice movement, activists and scholars have looked to other theories of justice to more fully explore and explain cases of injustice.

Environmental justice: 'beyond distribution'?

Several scholars have attempted to capture a definition of environmental justice which joins distributive concerns with other factors, but it is arguably Schlosberg (2007) who articulates the most comprehensive explanation. He characterises environmental justice as more than just fair distribution, noting that it is also concerned with how that distribution is shaped by: procedural fairness and parity in opportunities for participation in environmental decision making (procedural justice); the respect and recognition of various parties (recognition); and the capabilities of individuals and communities to function and flourish (capabilities). These factors overlap and are complex in themselves; the subsections below explore each one further.

Procedural justice

Procedural justice may necessitate a right to participate, to have one's interests included or the right to be represented (Bryner 2002: 45). Demands for procedural justice have been evident in claims for environmental justice throughout the history of the environmental justice movement, with scholars recognising that the rules and procedures governing how decisions are made are just as critical to perceptions of justice as the outcomes of decisions (Young 1990: 23; Schlosberg 2007). A lack of procedural justice can produce distributive injustice (and thus serves as an explanation for injustice under Walker's 'process' element of claim-making, see Figure 1.1) but can also, in itself, be a source of injustice (Preston 2015b; Walker 2012; Shrader-Frechette 2002). Over the last two decades, the importance of procedural justice in the environmental sphere has been recognised in international instruments, including: Principle 10 of the *Rio Declaration*; the *United Nations Economic Commission for Europe (UNECE) Convention on Access to Information, Public Participation in Decision Making and Access to Justice in Environmental Matters* (the *Aarhus Convention*); and, most recently, the United Nations Environment Programme's *Guidelines for the Development of National Legislation on Access to Information, Public Participation and Access to Justice in Environmental Matters*. Note the centrality of 'participation' in the titles of these instruments.

Procedural fairness and access to public participation can promote justice in several ways. As Hunold and Young note, it is '*prima facie* unjust to impose a risk on citizens without their having participated in the siting process' (1998: 83). Moreover, participatory processes respect the interests and autonomy of citizens, and are more likely to arrive at a decision that has considered consequences and alternatives, and is thus more distributively fair (1998: 87). The perception of fair processes also increases the likelihood that people will accept decisions even when the outcome is not favourable to them (Simcock 2014; Gross 2014; Skitka, Winquist and Hutchinson 2003; Bryner 2002; Tyler 2000). In some cases, the perception is critical; for example, after seven studies of water allocation scenarios with a range of stakeholders over ten years, Syme et al. (1999: 67) consistently found that fair processes were of 'paramount importance' to community acceptance of decisions. In other words, it is an expectation of *procedural* fairness, rather than substantive or outcome fairness, that underpins this dimension of environmental justice.

Several scholars have examined the question of ideal criteria for procedural justice, with the following key standards identified:

- **Notice:** Participation should be representative, which requires consideration of the diverse groups that may be affected by, and have an interest in, a proposal (Maguire and Lind 2003). These groups need to be informed of upcoming decisions involving the environment, and to have access to information relevant to the decision making process (Millner 2011). Notification and timing should be undertaken in ways that are perceived to be adequate and fair by interest groups; including clear, targeted and early notification, in line with the level of interest a group is likely to have. In each of the case studies in this book, there are examples of inadequate notice being perceived as an environmental injustice; for example: in the Bulga case (Chapter 3), the community complained that they were given minimal notice of a development expansion application and provided with insufficient time to make submissions; and in the Marcellus Shale case study (Chapter 6), similar concerns were raised about the notice and submission periods for gas well permit applications and the community's ability to respond within the limited time frames stipulated.
- **Information:** Information about environmental decisions must be affordable, effective, timely, adequate and representative of the public's concerns (Hunold and Young 1998; Smith and McDonough 2001; Gross 2007). The public needs to be able to access information throughout the decision making process in order to understand 'why and how' decisions are made. This was not the case in the Namoi catchment case study (Chapter 5), in which the lack of information available to the community during the assessment of exploration activities caused considerable disquiet. Information should also be presented in a manner that is transparent about the risks associated with land use change (Hunold and Young 1998), time should be allowed for the public to consider the complex information and issues (Maguire and Lind 2003), and technical material needs to be presented in straightforward language. When financial barriers exist to information access, assistance should be provided to correct disparities (Hunold and Young 1998). Access to information may be facilitated by freedom of information (FOI) laws and policies on government disclosure (Millner 2011) that include the need to use plain language. A frequent complaint by stakeholders interviewed for the case studies in this book was the difficulty in digesting the volume of technical information required in order to participate in decision making. Community members also complained of the costs of obtaining independent technical advice.
- **Consultation:** The public must be consulted on decisions in appropriate ways, such as holding meetings at convenient times, locations and frequencies (Smith and McDonough 2001), and using engagement methods and mediums to suit people's needs. Consultation must be multi-directional,

allowing participants to have a 'voice' through the ability to ask questions, express their opinions freely and be heard (Gross 2007), and decision makers should make genuine efforts to address their concerns (Millner 2011). Further, decision making processes must anticipate that consultation will allow for full participation in the process (Gross 2007) and take place at all phases over time, from formulation, to implementation and evaluation. Once-off public hearing style consultation is 'sporadic' and falls short of multi-phase discussion (Hunold and Young 1998). The case studies from NSW (Chapters 3 to 5) provide insights into the negative consequences of such once-off public hearings. Opportunities to participate in 'non-legislative and informal processes' connected to decision making are also critical to robust consultation. This can include, for example, community education, as well as the ability to participate in public debate and protests (Arcioni and Mitchell 2005). Later in this book, we will see attempts to constrain the right to protest by legislative intervention against coal seam gas exploration (Chapter 5) perceived as an injustice.

- **Decision making:** Participation in decision making must move beyond mere consultation with all stakeholders having the opportunity to participate in the decision making. Public input into the preparation of legally binding rules should be considered (Hunold and Young 1998); as we will see later, this was not the case in the rulemaking process concerning the extension of coal mining operations in the Liverpool Plains case study (Chapter 4). The process purportedly enabled public participation but, ultimately, ignored citizen contributions. Similarly, a complaint by stakeholders concerning shale gas extraction from the Marcellus Shale (Chapter 6) related to the narrow opportunities for participation in the drafting of Pennsylvania's new oil and gas law, and to concerns about the composition of regulatory advisory bodies and the influence of industry in the rulemaking process. As well as being able to participate in decision making, the decisions made by the participants in the process should be enforceable, authoritative and binding on public officials, with no party able to claim exclusive decision making power (Hunold and Young 1998). Decisions should also be transparent, impartial, justified and able to be corrected, including being modified or reversed in response to errors, oversights and new information (Smith and McDonough 2001; Gross 2007; Bell 2014). In the Bulga case study (Chapter 3), much of the community's complaint about the development assessment process centred upon the government's rejection of a decision from an impartial court and the legislative manoeuvres used to facilitate development.

- **Review:** Members of the public should have access to an independent and impartial body to challenge the legality of decisions, acts or omissions which affect the environment (Hunold and Young 1998). As Preston (2015b) highlights, the right to seek merits review, enabling independent expert review of the impact of an environmental decision, as well as judicial review, is critical. Further, information should be made available concerning the operation of courts and other relevant bodies with respect to environmental decisions (Millner 2011). In both the Namoi catchment case studies (Chapters 4 and 5), access to review was restricted through bureaucratic categorisations of development, and the defunding of public interest environmental legal services.

- **Standing:** Because it might be difficult to precisely identify who may be impacted by an environmental decision, standing to bring proceedings in courts and tribunals can be difficult to determine. This creates a barrier to justice. Interpretation of standing in proceedings should be broad to ensure effective access to justice (Millner 2011). In each of the case studies in this book, issues of standing constrained the opportunities for parties to seek review of decisions. In the Marcellus Shale case study (Chapter 6), in particular, several debates about standing were heard in the courts to determine who could legitimately challenge the state's new oil and gas law.

Some scholars have argued that the existence of participatory opportunities is not sufficient to ensure justice and that justice requires the satisfaction of substantive standards as well (Bryner 2002). The measurement of such standards, however, is incredibly difficult, leaving the existence of procedural norms as the main indicator of successful environmental law and policy. While it is self-evident to claim that the possession of a right is separate from the ability to enforce it, it is a reality that any number of barriers can limit or exclude public participation. As noted, efforts to deal with geographic, economic or educational constraints can be accommodated within certain procedural conditions but, in practice, these often do not go far enough to ensure recognition of non-dominant perspectives, or fair and respectful treatment of all stakeholders (Schlosberg 2007; Young 1990; Fraser 1998). Lake argues that many scholars have an 'unnecessarily truncated notion of procedural justice' and have overlooked the need for democratic self-determination and local community control (1996: 162). As he puts it, the problem of unequal distribution:

> [D]oes not arise simply because marginalized communities lack the power to influence the location of environmental problems. It derives, instead, from the inability of such communities to influence the process that produces environmental problems in need of distribution.
>
> (Lake 1996: 171)

Measuring justice by the existence of procedural opportunities may well be illusory; justice also requires 'fundamental changes in the way in which economic power and political power are distributed' (Bryner 2002: 46). Social misrecognition must, therefore, also be addressed to improve parity of participation. Young's (1990) and Fraser's (1998, 2000) research on this subject within the discipline of political science have brought the concept of recognition to the fore as critical to improving participation and necessary to correct distributive injustices. Schlosberg (2007) incorporates the concept of recognition within his conceptualisation of environmental justice, accepting, in particular, claims that seek respect and recognition of non-dominant perspectives.

Justice as recognition

To differing extents, environmental injustices are defined by inequities of relative political power, opportunity and outcome:

> Environmental injustices and disparities in the benefits that flow from natural resources development are a manifestation of more fundamental problems of poverty and political and economic disadvantage, democratic inequality and lack of participation, and other manifestations of social injustices and unfairness.
>
> (Bryner 2002: 47)

Some justice theorists (such as Miller) maintain that recognition is a precondition for procedural justice in that if 'procedural justice . . . is attained, recognition is included and so is to be assumed' (Schlosberg 2007: 26). Empirical claims for environmental justice suggest, however, that the connection between recognition and participation is far more complex. A lack of social, cultural and political recognition has long been identified as a barrier to participation in the political institutions concerning public decision making. Arnstein's (1969) 'ladder of citizen participation' elaborates a typology of levels of citizen participation, from non-participation and tokenistic methods, through to delegated power and citizen control, linking more authentic forms of public participation with recognition.

Young (1990) argues that a focus solely on the allocation of goods fails to illuminate the social, political and institutional contexts in which maldistribution takes place. Maldistribution arises from a lack of recognition of difference; that is, the ways in which 'some groups are privileged while others are oppressed' (Young 1990: 3). Individual and cultural misrecognition in social and political spheres not only produces distributive injustice but also serves as a source of injustice in its own right through the disrespect and devaluation of non-dominant perspectives (Young 1990; Fraser 1998, 2000; Honneth 1995). As Foreman argues, claims for environmental justice are 'mostly about accountability and political power', and are 'often anchored . . . in a desire for transformed power relationships to be achieved' (1998: 58–59). Contemporary conceptions of environmental justice have thus become increasingly concerned with politics, power and the ability for minority groups to be recognised.

There is some debate in the literature concerning 'recognition'. Honneth (1995: 132–134) frames recognition as a question of personal identity, with self-worth a matter of psychological interpretation of one's approval and acceptance by others. He identifies three types of misrecognition: bodily violation (e.g. physical abuse), the denial of equal rights, and the denigration of individual or cultural ways. He claims that individual perceptions of these actions as forms of disrespect cause humiliation and damage to self-identity. In her analysis of fairness and justice in environmental decision making, Gross (2014) details how interactional justice is critical to participant perceptions of justice. She draws particularly on the work of Bies (2001) and Tyler and Blader (2003) to argue that the way people are treated can be distinct from procedural decision making functions, and can have a powerful impact upon perceptions of fairness and justice. Injustice within the interpersonal or interactional aspects of procedures can give rise to 'intense and personal pain' that is 'experienced as a profound harm to one's psyche and identity – that is, one's sense of self' (Bies 2001: 90). Tyler (2000) suggests that, for decision making processes to be viewed as fair, the neutrality and trustworthiness of authorities, and the treatment of stakeholders with dignity and respect are just as important as the existence of opportunities to participate. Through the case studies in this book, we see examples where people's perceptions of poor treatment from decision makers can give rise to a sense of interactional injustice. For instance, in the Bulga case study (Chapter 3), community members' requests to meet with politicians were repeatedly denied, while industry representatives were able to have private meetings with senior ministers.

Fraser alternatively anchors her definition of recognition in the social order of individuals and groups, defining structural misrecognition as a 'status injury' (1998: 25). Injustice, as Fraser argues, is 'rooted in patterns of representation, interpretation, and communication' (1998: 7). This necessitates an exploration of not only social and political structures and practices, but also the norms and rules that govern them and the 'language and symbols that mediate social interactions within them' (Young 1990: 22). Fraser notes that these status-based structural injustices can include:

> [C]ultural domination (being subjected to patterns of interpretation and communication that are associated with another culture and are alien and/or hostile to one's own); nonrecognition (being rendered invisible via the authoritative representational, communicative, and interpretative practices of one's culture); and disrespect (being routinely maligned or disparaged in stereotypic public cultural representations and/or in everyday life interactions).
>
> (Fraser 1998: 7)

Schlosberg argues that, while this dichotomisation of definitions of recognition is valid, the concepts of 'identity/psychology' and 'status' should not be mutually exclusive in constructing a theory of misrecognition; misrecognition can be simultaneously experienced individually and structurally (2007: 20). In this sense, the inclusion of 'recognition' in conceptualising environmental justice represents a significant departure from purely distributional approaches. Recognition is not perceived as a 'good'; rather, it is 'a precondition of membership in the political community' (Schlosberg 2007: 23). When a previously non-recognised group is attributed appropriate recognition, it is not at the expense of another group; in other words, recognition is not a zero–sum game, a factor that a purely distributional frame fails to capture.

Empirical accounts cite ample evidence of social and cultural misrecognition in environmental decision making (Schlosberg 2007: 58–64). In the environmental sphere, misrecognition can occur when there is a failure to take into account the particular knowledge, experience and needs of racial, cultural or socio-economic groups. Structural inequities are embedded when minority groups are 'devalued and/or ignored' (Schlosberg 2007: 61) through the imposition of dominant cultural standards. Hillman (2006) provides an example when describing how colonists dispossessed and excluded Indigenous people from their understanding and consequent management of the land in the Hunter Valley in NSW, Australia. The colonists did not appreciate the long history of adaptive management by Indigenous people, nor the local environmental diversity and variability. The colonists saw the land as pristine parkland suited for settlement. This illustrated a failure to recognise special circumstances which translated into structural inequity, and consequently 'left a legacy of misunderstanding and environmental degradation that has continued to promote procedural and ecological injustice' (Hillman 2006: 698).

The dominance of an 'environmental economics' discourse to justify and legitimise 'mutual benefit arising from pricing and commodifying nature' is another example of the reinforcement of status hierarchies (Martin, McGuire and Sullivan 2013: 124). This is apparent in the case studies presented in this book which provide examples of industry and government prioritising the wider economic benefits of development while downplaying local environmental harm and social impacts. The case studies trace how this status hierarchy is reinforced. It is especially manifest in the preference for 'hard' scientific data over 'soft' citizen evidence and place-based values in decision making processes. In the Bulga case study (Chapter 3), for example, the reliance on a cost–benefit analysis to capture the costs and benefits of extraction discounted the significance of social and environmental impacts.

Disrespect and malrecognition, as another form of status injustice, includes acts of 'cultural domination', which not only injure social status but constrain participation (Preston 2015b: 39). This may be evident in the derogatory labelling of individuals and groups who may be opposed to development as 'NIMBYs' (or other pejorative terms). In the Marcellus Shale case study (Chapter 6) there are examples of derogatory terms used to describe those opposed to shale gas development. In some cases, those engaged in protest activity were subject to government surveillance and listed as potential terrorist threats. This impacted their social status and had a chilling effect upon participation. Similarly, those who protested against gas extraction in the Pilliga (Chapter 5) and legitimately challenged development through the courts were labelled by industry and government as 'vigilante green groups' who were attempting to sabotage economic development.

The use of punitive and 'malignant' legal actions (such as strategic litigation against public participation, or the threat of adverse costs orders following litigation), which are instigated to prevent or constrain participation, can also amount to malrecognition (Arcioni and Mitchell 2005; Millner 2011; Preston 2015b). Such malrecognition was evident in the Bulga case (Chapter 3), in which the government initiated regulatory change to effectively reverse an independent environmental court's refusal of a coal mine development. The legislative amendments were a response to the community's successful challenge to development consent. The removal of government funding for public interest legal advice, which also occurred in the Bulga case, provides another example of action designed to impair participation.

Although recognition is a relevant factor within the concept of environmental justice, it, too, 'can only go so far' (Schlosberg 2012: 452). Justice requires that recognition be transformed into concrete processes which provide actors with 'some locus of control over their destinies as part of a recognition of identity and place' (Adger, Barnett, Chapin and Ellemor 2011: 21, as cited in Schlosberg 2012: 452). The sense of injustice expressed in real-world claims for environmental justice often includes such matters as the freedom and well-being of individuals and communities, extending the concept of environmental justice beyond respect and inclusion into a consideration of capabilities.

Capabilities

The capabilities approach provides a space for considering the numerous interlinked justice concerns in claims of environmental inequality (Schlosberg 2007). Sen (1993) originally outlined the capabilities approach as a means to understand social inequalities by assessing individuals' well-being according to the capabilities they possess, and their freedom to function in the manner they desire. The capabilities approach is concerned with evaluating a 'person's advantage . . . in terms of his or her actual ability to achieve various valuable functionings as a part of living' (Sen 1993: 30).

Within the environmental justice sphere, scholars have increasingly turned to the capability approach to conceptualise environmental inequalities, to evaluate environmental law and policy, and to examine cases of environmental conflict (Schlosberg 2007; Holland 2014; Tschakert 2009; Schlosberg and Carruthers 2010; Roesler 2011; Schlosberg 2013; Griewald and Rauschmayer 2014; Groves 2015). Schlosberg, in particular, has advocated for the inclusion of capabilities in conceptualising environmental justice. He maintains that equitable distribution cannot occur without recognising the relationship between public participation, recognition and broader community capabilities. For Schlosberg, the capabilities framework offers a multidimensional approach which is more comprehensive than exploring issues of recognition and participation as discrete components of environmental justice. Walker (2012) maintains that the capabilities framework is integrative and inclusive of other justice elements, bringing together distribution, recognition and participation to explore threats to capabilities to function. As Tschakert suggests, the capabilities approach 'characterizes the plurality and multiple spaces of social and environmental justice' (2009: 709).

Capabilities are things which enable 'valuable functionings'; that is, the inter-related combinations of valuable 'beings and doings' (such as reading, working, being well nourished, being healthy, being safe, being part of a community, being respected) (Sen 1993: 31; Robeyns 2005; Alkire 2005). For example, if reading is a valued functioning, then education and literacy would be the necessary capabilities for that functioning (Schlosberg 2007). As Robeyns explains, capabilities are potential functionings and, taken together, all of an individual's capabilities constitute their capability set, which represents their 'real or substantive freedom' to do and be what they want (2003: 544). Central to measuring capabilities is whether or not individuals possess the freedom to undertake and achieve these doings and beings – the *ability* to read, the *ability* to work, the *ability* to be well nourished, and so on (Robeyns 2003; Nussbaum 2011). The concern is, thus, with the *real opportunity* to do something rather than only with functional outcomes. Freedoms or opportunities which are held legally or theoretically but which lie outside a person's reach (for financial or geographic reasons, for example) would not be part of an individual's capability set (Alkire 2005). That is, capability combines functioning *and* freedom. The capability approach proposes that social arrangements 'should be evaluated according to the extent of freedom people have to promote or achieve functionings they value' (Alkire 2005: 122).

Capabilities are determined by a range of factors. Financial resources are one input into achieving a particular functioning, but other circumstances will also be relevant (Robeyns 2005). This might include – but is not limited to – individual or personal factors, such as physical abilities and characteristics; social factors, such as social and legal norms, power relations, hierarchies or gender roles; and environmental factors, such as geographical location and climate (Robeyns 2005). A capabilities analysis directs us to look at not only what goods and resources are available, but also how they combine with various circumstantial factors and are converted into the capabilities that enable individuals and groups to function as they desire.

Determining what capabilities are relevant to achieve particular functionings can form a metric for evaluating whether or not particular laws or policies expand freedoms (Robeyns 2005; Roesler 2011). There has been much debate over what constitutes relevant capabilities. Sen notes five essential freedoms in *Development as Freedom* (1999: 10): political freedoms, economic facilities, social opportunities, transparency guarantees and protective security. However, he remains elusive as to the particular capabilities necessary to enable valuable functionings and resists specifying a set list. For Sen, a single predetermined list of capabilities generated by the theorist is problematic because capability assessments may be undertaken for a variety of purposes, and it is essential that people can choose via democratic processes which capabilities apply to themselves (Sen 2005). Nussbaum argues there is a need to define a core set of capabilities, which she lists as: life; bodily health; bodily integrity; senses, imagination and thought; emotions; practical reason; affiliation; other species; play; and control over one's environment (both material and political rights) (2000: 78–80; 2006: 76–78). Ballet, Koffi and Pelenc (2013), however, argue that the capabilities approach does not necessarily need to have a normative criterion of requisite capabilities so long as it can lead to judgements between alternative policies or situations. In this way, the capabilities approach may be considered as 'between' descriptive and normative approaches – one that 'permits an evaluative judgment but not a prescriptive judgment' (Ballet et al. 2013: 30). As Robeyns suggests, the capabilities approach is 'primarily and mainly a framework of thought, a mode of thinking about normative issues; hence a paradigm – loosely defined – that can be used for a wide range of evaluative purposes' (2005: 96). The definition of an ideal state of justice is not required, since the goal is to determine whether a particular policy alternative improves or reduces capabilities. In this sense, the assessment of injustice does not require the a priori determination of a capability set; injustice occurs where there is, quite simply, 'a reduction of the capabilities space' (Ballet et al. 2013: 31).

Under the capabilities approach, the measure of justice in social arrangements, such as laws and policies, is in whether or not these institutions expand or contract people's capabilities to function: 'What is ultimately important is that people have the freedoms or valuable opportunities (capabilities) to lead the kind of lives they want to lead, to do what they want to do and be the person they want to be' (Robeyns 2005: 95). The capabilities approach holds that social arrangements, such as laws and policies, should improve the quality of life of people by increasing their opportunities to achieve what they consider to be important (Nussbaum 2011; Sen 1999). Injustice occurs when a capability is restricted or denied (Schlosberg 2007).

Schlosberg (2012) explains that a capabilities approach can more readily encompass the human–nature relationship by incorporating the environment in two main ways: through recognition of the environmental conditions that underpin human capabilities; and by applying the capabilities approach to the environment itself, particularly the functioning of ecological systems. With regard to the former, a sustainable environment is necessary to enable freedoms for current and future generations and has, therefore, been described as a 'meta-capability' because all human capabilities depend upon it (Schlosberg 2012). As for the functioning of nature itself, Schlosberg (2012) indicates that applying the capabilities approach is more complex because it highlights conflicts between fulfilling both human and environmental capabilities; nonetheless, the process of determining and prioritising environmental threats to human and community capabilities, such as climate change, exposes the impact that humans have on the environment and the status of ecological systems. Schlosberg summarises:

> If the capabilities approach is about functioning, and we all need particular aspects of the environment to help us function, functioning for human beings means acknowledging the human dependence on environment, and providing for those ecological support systems that make that functioning possible.
>
> (Schlosberg 2012: 456)

Two recent empirical analyses, one by Groves (2015) and one by Griewald and Rauschmayer (2014), of capabilities and environmental justice are worth noting at this juncture. Exploring the governance processes relevant to the construction of the South Wales Gas Pipeline in the UK, Groves (2015) notes how the failure to recognise place attachment can restrict capabilities, resulting in environmental injustice. As discussed in Chapter 1, place attachment is the social and emotional connection between people and places. It is comprised of physical characteristics, such as the material environment and the 'dispositions, attitudes, beliefs and practices that are embedded within that environment' (Groves 2015: 856). Place attachment is dynamic, linking 'place, identity and agency' to provide a basis for individuals and groups to navigate uncertainties and shape their future:

> Attachment is a process through which an uncertain future is tamed and made liveable in the present through the creation of affective bonds of trust and attendant expectations about how the world should be. Attachments are thus a way of giving shape to the future that encourage particular ways of acting in the present – strategies for living with uncertainty.
>
> (Groves 2015: 868)

Actual and perceived changes to the characteristics of a landscape can erode place attachments by imposing a new and uncertain future, destabilising the identity and agency of individuals and groups (Devine-Wright 2011a; Lin and Lockwood 2014; Groves 2015). Groves characterises place attachment as a capability which, when denied, impairs self-definition and self-determination (2015: 854). Groves interviewed community members opposed to the construction of the South Wales Gas Pipeline and found that both the threatened and actual disruption to the physical environment from the pipeline construction and the relevant project approval process had a negative impact on their place attachment. In particular, the narrow parameters to express complex concerns about place transformation and to challenge the project failed to recognise the identity of citizens as embedded in place, and also restricted their agency. Attachment became 'colonised' as objectors struggled to 'translate values into the language of the colonists' – a requirement for them to participate in the development assessment process (Groves 2015: 870). Ultimately, this restricted the capability of the community to influence their future, and constituted an environmental injustice.

We noted earlier that the dominance of economic concerns as a basis for development assessment can lead to a lack of recognition of place-based values. In light of Groves's (2015) study, it is evident that this lack of recognition of place-based concerns can also restrict capabilities. This is apparent throughout the case studies discussed in this book, in which communities complain not only of their inability to effectively articulate place-based concerns within decision making frameworks, but also how the lack of recognition disrupts their capability to exercise control over their local environment and their future.

Griewald and Rauschmayer (2014) adopt the capability approach to analyse a public dispute over tree clearing in the German state of Saxony. They set out to uncover why certain actors (such as the government and NGO) took the actions that they did throughout the conflict. The dispute concerned the removal of 6,500 trees in Leipzig from a floodplain forest by the state dam authority, following a decree from the federal state ministry which permitted the cutting down of trees on and along dykes as a flood protection measure. The state dam authority, responsible for major water bodies (which includes dykes), initiated the tree clearance without seeking public input. The local population and environmental NGO strongly criticised the tree clearance but the city administration, responsible for smaller water bodies within the city boundaries, did not object to the clearance measures.

Griewald and Rauschmayer (2014) focus on the factors that influenced the freedom and agency of the actors in the conflict. They found that the dam administration benefited from a broad legal framework (the decree) that enabled urgent action in the face of 'imminent danger' (the exposure of the floodplain location). The administrators also benefited from political support at local and state levels for technical (rather than nature-based) approaches to floodplain management. In terms of capabilities, they were able to act swiftly and autonomously in implementing the tree clearance. The city administration had the agency to intercede within the existing legal framework, particularly given that another city administration elsewhere had intervened in similar circumstances, yet they did not do so. This decision directly impacted upon the capabilities of the NGO and, ultimately, its achieved functionings, resulting in environmental injustice. The NGO's only remaining capability was to bring a lawsuit against the dam administration.

As Griewald and Rauschmayer (2014) explain, a lack of recognition can constrain the capabilities that enable certain desired functionings (e.g. participation in environmental decision making). Throughout the case study chapters in this book, similar factors are evident. For example, the defunding of public interest environmental law services (in the Bulga case in Chapter 3) is an example of malrecognition and of limiting the capacity of the public to participate.

An important point about the Groves (2015) and Griewald and Rauschmayer (2014) studies, which is consistent with the overall approach advocated by Schlosberg (2007), is the shift from an individualistic unit of analysis to a collective one. This shift was necessary given that the key actors in both studies were collective entities and many of the justice issues expressed were shared and relational. Robeyns (2006) argues that the capabilities approach is neither methodologically nor ontologically individualistic, even though the literature on capabilities is largely focused on justice to individuals. Rather, the capability approach accounts for social relations and the impact of social structures and institutions on individuals and is, therefore, a relevant approach for examining group interactions. Schlosberg notes that injustice can be experienced by groups and can be explored at that level (2007: 35). Revisiting Sen's work, Schlosberg sees significance in his insistence upon public deliberation in determining capabilities; communities name their own capabilities that are relevant to their functioning (2007: 37). It is groups and communities who suffer disproportionate environmental outcomes, and it is groups and communities who demand justice in terms of recognition, participation and overall capabilities in a collective sense (Roesler 2011).

Exploring claims of environmental injustice through the lens of capabilities enables us to see how concerns about distribution, participation and recognition are 'thoroughly tied'; they are all 'components of a more broad set of factors necessary for our lives to function' (Schlosberg 2007: 33–34). The overall measure of injustice is whether the capabilities deemed necessary to function in particular ways have been constrained. We will see these concerns emerge throughout the case studies in this book, ranging from specific capabilities, such as the ability to access information, through to broader and complex capabilities, such as the ability for community identity and culture to survive in the wake of land use decisions.

A plural concept of environmental justice

The concept of environmental justice has evolved from a relatively narrow base of distributive justice to incorporate the additional frames of procedural fairness, recognition and capabilities. While some have expressed concern with moving 'beyond distribution' (Getches and Pellow 2002), the alternative approaches to environmental justice discussed above are not exclusionary, and can be situated within a pluralistic framework. Schlosberg maintains that the articulation of environmental justice as interlinked and overlapping elements does not 'dismiss distribution, or to call for a move *beyond* distribution'; rather, it puts distribution 'in its place alongside other components of a comprehensive understanding of justice' (2007: 12). A plural theory of justice can 'offer a satisfying account of all the phenomena within its domain' (Bohman 2007: 268). It may court conflict, but this can be productive. It is important to keep in mind that theories of justice 'are not arbiters of disputes about difficult questions . . . [t]hey are general interpretive frameworks rather than complete rankings of principles and values and independent standards of judgments' (Bohman 2007: 274–275). A theory of justice must, therefore, be guided by the reality of how it is experienced in practice. A plural theory of environmental justice is necessary since pluralism is 'both empirically real in the expression of justice claims, and pragmatically necessary to avoid the mistakes of exclusion' (Schlosberg 2007: 167).

In linking distribution, procedure, recognition and capabilities, Schlosberg's definition extends environmental justice beyond the geographic distribution of environmental risks and harms to explore the real impact upon individuals and communities. He notes, '[t]he focus is not simply on a conception of distribution, or of recognition . . . but more holistically on the importance of individuals functioning within a base of a minimal distribution of goods, social and political recognition, political participation, and other capabilities' (Schlosberg 2007: 34). Ultimately, his definition of environmental justice 'offers a broad and inclusive definition' of justice, one which focuses not just on the distribution of goods but also on 'how those goods are transformed into the capacity for individuals to flourish' (Schlosberg and Carruthers 2010: 15).

Schlosberg's multivalent conceptualisation of environmental justice is used throughout this book to provide a normative framework for evaluating perceptions of justice in land use decision making. His capabilities-based definition adds depth to theories of recognition and procedural fairness by bringing a freedom and agency lens to explore and assess situations of injustice. Through a concern for 'real-world consequences and alternatives, rather than ideal institutions', a capabilities-based definition of environmental justice enables a focus on 'what people can actually do and be as a result of environmental decisions' (Roesler 2011: 73). As a basis for assessing environmental policy, a capabilities-based conceptualisation of environmental justice enables a broad understanding of well-being that accounts for the non-economic value of the environment to human life (Holland 2014). Moreover, through this broad notion of well-being, a capabilities-based approach is better placed to reveal the inequities that are not visible in an economic analysis of environmental policy (Holland 2014: 63–64). This enables a deeper understanding of actor's freedoms and agency and why conflict takes a particular path, which may prove useful in resolving conflict or providing an ethical basis for future policy interventions (Groves 2015; Griewald and Rauschmayer 2014; Roesler 2011).

Causes of environmental injustice

The foregoing sections were concerned with developing the normative concept of 'justice' as one element of environmental justice claim-making (see Figure 1.1). In this section, we turn to the 'process' element of Walker's (2012) framework, to better understand how and why environmental injustices are produced. There is, as Walker notes, 'a degree of overlap between justice concepts and questions of process' (2012: 65), and some of the discussion in the previous sections alludes to process claims (such as the role of political power in producing misrecognition). Returning to Schlosberg's multivalent definition of environmental justice, Walker reiterates that distribution, participation and recognition are not only forms of injustice but also distinct *causes* of injustice:

> [I]f you are not recognized you do not participate; if you do not participate you are not recognized. In this respect justice must focus on the political process as a way to address both the inequitable distribution of social goods and the conditions undermining social recognition.
>
> (Schlosberg 2007: 26, cited in Walker 2012: 65)

Understanding the causes of environmental injustice is important to many activists and scholars, particularly to support effective efforts to address injustice (Bell 2014; Osborne 2015). Brehm and Pellow argue that environmental justice issues are fundamentally social problems and to frame them as ecological problems 'runs the risk of missing the point that ecological violence is first and foremost a form of social violence, driven by and legitimated by social structures and discourses' (2013: 308).

The causes of environmental injustice have been broadly categorised under racial discrimination, economic explanations and socio-political explanations (Brehm and Pellow 2013; see also Walker 2012).

Racial discrimination was the initial focus of the environmental justice movement and continues to be a central cause of environmental injustice. Environmental justice emerged from activism and research in the US that identified that communities of colour and other racial minorities were disproportionately exposed to pollution and environmental risk, as a result of the siting of hazardous waste dumps and polluting facilities near their communities (United Church of Christ Commission for Racial Justice 1987; Bullard 1990; Pulido 1996). A quintessential example is the Warren County case, which involved toxic PCB-contaminated soil being disposed of near the 75 percent African-American Shocco Township (Szasz and Meuser 1997). This dumping was approved by the US Environmental Protection Agency (EPA). The political movement that resulted from this unequivocally discriminatory action was one of the first in broader environmental justice activism. Catalysed by this issue, the discriminatory distribution of environmental harms became the focus of much environmental justice research, which has demonstrated the link between exposure to environmental hazards and risks and race, powerfully articulating the impacts of exposure (e.g. Bullard 1990; Taylor 1997; Newell 2005). Scholars have argued that environmental racism is a form of institutionalised discrimination that operates through unequal power arrangements, particularly where 'ethnic or racial groups are a political and/or numerical minority' (Bullard 2002: 35).

Economic explanations demonstrate environmental injustices resulting from paradigms, decisions and actions primarily within contemporary capitalism (Bell 2014; Walker 2012). Contemporary capitalism is often linked to environmental injustice because of the way assumptions about the benefits of the market system influence society's perception of the environment as separate from humans and the market (Bell 2014). Nadeau (2006) argues that neoclassical economics and environmental sustainability are incompatible because the neoclassical economic paradigm views the market as a closed system that seeks and, when functioning well, achieves, perpetual economic growth. In this paradigm, the environment is commoditised: 'it has value only as environmental goods, services, and amenities that can be bought, sold, traded, saved, or invested, like any other commodity' (Nadeau 2006: 125). Private actors make decisions in the market system to maximise their utility, which leads to optimal social outcomes for the environment, assuming that the environment has a price or value in the market system (Nadeau 2006). However, many aspects of the environment are not priced in market systems, including ecological services and environmental impacts such as environmental degradation and pollution. As a result, significant environmental degradation, waste and pollution are products of economic ideologies.

Socio-political explanations include analyses of politics, power and culture. Scholars have drawn upon theories of political ecology and, in particular, urban–political ecology to foreground the causal relationships between politics and social power relations, and situations of environmental injustice (Andrews and McCarthy 2014; Walker 2012; Heynen et al. 2006). By drawing attention to the 'informal, extra-legal, and tacit relationships and dynamics' between people and environments, power imbalances and their consequences may be identified (Andrews and McCarthy 2014: 8). As Harvey argues, 'much of what happens in the environment today is highly dependent upon capitalist behaviors, institutions, activities, and power structures' (1996: 196), rendering critical an account of the actors and relationships that produce this power (Heynen et al. 2006).

Central to each of these explanatory frameworks is the unequal distribution of power. Power may be structural (referring to political or economic frameworks), material (object-based) or discursive (language-based), and operates to exclude certain interests from decision making processes (Walker 2012: 72; Heynen et al. 2006). Relations of power however may be difficult to detect and understand; theories of power provide an opportunity to understand the often-invisible processes of oppression, helping to problematise the processes and enabling the incorporation of other knowledges (Osborne 2015).

In his study of coal mining in Central Appalachia's Clear Fork Valley, Gaventa (1980) illustrates how theories of power may be used to illuminate injustice. The purpose of his research was to understand the 'quiescence' or non-rebellion of miners against their inequality and exploitation, and local environmental harm. Gaventa traces the story of the 'colonization' of the area, from the company's acquisition of the land; to their control of local economic facilities upon which the miners depended; to their co-option of local government, professionals and merchants who acted as agents for the company; and, finally, to the company's devaluation of the local culture and way of life. He questions why (notwithstanding some instances of local protest) there was largely a state of 'non-rebellion' against this inequality. Gaventa argues that the persistent quiescence was a product of, and maintained by, the disparate power relations between the miners, on the one hand, and the multinational owners of the mining company and the local elites, on the other.

Adopting a 'three dimensional' view of power, Gaventa details how the company, the government and their local agents curtailed attempts at resistance and entrenched a culture of quiescence through both obvious and hidden acts of power. The first dimension or 'face' of power refers to observable power, where 'A has power over B to the extent that he can get B to do something that B would not otherwise do' (Dahl 1957: 202–203). Power can work directly to limit participation in this way and to limit the scope of the political process to non-controversial issues or to exclude certain participants. The second dimension of power focuses on 'nondecision-making', in which demands for change are silenced before they are voiced, through the shaping of values and institutions that control what is debated and what is not (Bachrach and Baratz 1970: 44). Schattschneider succinctly observes that this second face of power is apparent when '[s]ome issues are organised into politics while others are organised out' (1960: 71), and when cultural hegemony is declared and justified in a 'mobilization of bias'.

Even if the 'rules of the game' (Bachrach and Baratz 1970) fail to frustrate attempts to challenge those in power, power is also evident in a third, hidden dimension. This is conceptualised by Lukes (2005) as the capacity to influence the thoughts and expectations of others through the manipulation of ideology. The influence mechanisms include the symbols and language that 'construct and reinforce conceptions of who has power and who is powerless; the very development (not only the wielding) of a mobilization of bias' (Bernal 2011: 158). Lukes argues this form of power is most 'effective and insidious' when it is used to prevent conflict from arising at all (2005: 27). This third dimension, therefore, seeks to challenge the appearance of consensus by highlighting the shaping of 'perceptions, cognitions and preferences in such a way that they accept their role in the existing order of things' (Lukes 2005: 28).

Gaventa's research highlights the direct and indirect ways in which powerlessness may be established and upheld to perpetuate environmental injustice. His study demonstrates how all three dimensions of power combined to reinforce the authority of the company and their agents, and the powerlessness of the miners. The political might of the company over the miners was observable in a range of public contests in which grievances concerning labour and economic conditions were 'organised out' with the assistance of local elites and government. Ultimately, in the face of continual defeat, the miners internalised a sense of powerlessness and accepted their state of inequality, leading to their quiescence and failure to take action.

In their analysis of environmental justice in a payments for ecosystems services scheme in Rwanda, Martin et al. (2014) found that even the subtle exercise of broad discursive power resulted in non-recognition. As they note, discursive power can 'suppress the right to alternative ways of thinking', including the ways in which justice is framed – for example, when a conservation organisation holds itself out to be the 'legitimate source of justice norms' – which can, in turn, regulate the visibility of conflict and, ultimately, 'whose knowledge counts' (Martin et al. 2014: 169). Aligned with Gaventa's study, Martin et al. note that power may be 'embedded in cultural norms at large in society, in discourses that frame the ways in which problems and their solutions are conceived, and even in the structures of language itself' (2014: 169).

In conflicts over resource extraction, such power contests play out particularly over matters of scale; that is, in terms of the level at which projects are assessed and controlled, and at which the benefits and costs of extraction are articulated and distributed. At the very centre of extractive development disputes are often tensions between the 'fixity' of mineral deposits and the localised social and environmental impacts of extraction, and the 'multiscalar forces' that compete for their development (Huber and Emel 2009; Kurtz 2003). As Huber and Emel explain, this creates 'struggles over scale':

> The debates and conflicts are a matter of scale: participants deploy scalar terms and activate scalar commitments, and scale becomes institutionally materialized in ways that facilitate and pose barriers to the production of resource wealth.
>
> (Huber and Emel 2009: 371)

Scales, such as 'local', 'regional', 'state', 'national' and even 'international' have a material, biophysical component, but they are also discursive in that they are socially constructed and performed by actors to support their claims. In this sense, scale can operate as a 'framing device' used by those in power to establish the scale of a dispute in order shape and channel conflict towards their desired outcomes (Huber and Emel 2009: 371; Kurtz 2003; Bickerstaff and Ageyman 2009; Jessup 2013). As MacKinnon elucidates:

> [S]cales and scalar relations are shaped by the processes of struggle between powerful social actors and subaltern groups. The former seek to command 'higher' scales such as the global and national and strive to disempower the latter by confining them to 'lower' scales like the neighbourhood or locality.
>
> (MacKinnon 2011: 24)

Scale may be 'invoked strategically in order to build allies and coalitions, or to legitimate inclusion in decision making processes' (Jessup 2013: 79). Those seeking to advance development may attempt to engage 'higher' scales, such as global or national economic development or energy security, while simultaneously downplaying and confining concerns about the impacts of development to a 'lower' local scale (MacKinnon 2011: 24; Sica 2015). Contests over 'scale frames' can also give rise to 'spatial ambiguity', which can either be a hurdle for environmental justice advocates to overcome or a tool they can use to aid in advancing claims (Kurtz 2003: 913).

The law is an important mechanism of power that can both enable and constrain actors in negotiating the scale of disputes:

> In its enabling form, the law can serve as a vehicle for the recognition of actors and groups as political actors. It can also serve to legitimize their claims for justice, and, though this is more debatable, create the space for ameliorating these injustices through symbolic recognition and legitimation of a group's claims as just claims in violation of the law, or substantively through adjudication of a legal remedy to rectify said wrong. In contrast, in its role as a constraint, a judicial finding against an actor or group can delegitimize and silence their claims. Consequently, it can work to entrench and reproduce disparate power relations in a society.
>
> (Bernal 2011: 145)

Recent empirical analyses of land use conflict have explored the ways in which the law is used by those in power to structure the spaces and scales of disputes. Andrews and McCarthy (2014) explore how the law and extra-legal dynamics combined to undermine local control over shale gas development in Pennsylvania and to centralise state power. Their preliminary analysis is expanded upon in the Marcellus Shale case study (Chapter 6). Jessup's (2013) examination of the approval processes relevant to a waste facility in NSW, Australia, similarly highlights how the law may be used to construct and prioritise the regional scale over local impacts. In that case, powerful actors imposed an ideological notion of the 'public interest' at the regional level, which displaced local concerns about the impact of the development. Finally, in a recent analysis of aggregate extraction in Ontario, Canada, Van Wagner (2016a, 2016b) examined how legal structures privilege ownership and extractive land uses while simultaneously obscuring more-than-ownership relations with the land, such as place-based attachments.

Struggles over scale are evident in the case studies presented in this book. One of the most prominent complaints articulated by communities in the case studies was their inability to control the concentrated local impacts that had either already resulted from extractive development or would likely eventuate in the future. In each case study there was a sense that the laws and policies implemented to manage conflict constrained local control over development by prioritising different scales of assessment and benefit. The case studies trace how these scales were reconstructed, and how they were reinforced through other discursive acts (Chung and Xu 2015), for example, through the non-recognition of particular place attachments, or through limiting opportunities for public participation.

Disentangling these dynamics can reveal how power relationships evolve and play out in shaping how actors are enabled and constrained within conflicts over resource extraction. This resonates with Schlosberg's (2007) capabilities-based definition of environmental justice. His conceptualisation encourages us to explore whether actors (individual or collective) possess the freedom to function as they desire, or whether their capabilities are denied. Tracing how relationships of power affect capabilities can expose the causes of injustice. By focusing on these matters of process as well as the evidence of injustice, we can gain a deeper understanding of the experience of injustice in order to develop more effective solutions.

Conclusion

In this chapter, we have covered significant theoretical terrain to explain how environmental justice is conceptualised. A number of key themes discussed in this chapter can be taken forward which will prove useful in the analysis of conflict over extractive development in the subsequent case study chapters of this book.

The broad notion of environmental justice, as a normative concept of 'how things ought to be', is appropriate for understanding the multifaceted claims for justice which lie at the heart of conflicts over extractive development. The plural definition of environmental justice proposed by Schlosberg (2007), which brings together notions of distribution, participation, recognition and capabilities, offers a comprehensive framework to understand how individuals and communities might articulate perceptions of injustice in land use decision making, and to appreciate how a denial of justice in any of these realms might prompt conflict.

While we have parsed the various conceptual elements of environmental justice separately throughout this chapter for analytical convenience, in practice they interconnect and overlap. As Schlosberg explains:

> [W]hile a group may call for equal distribution of environmental risks, it will also complain about the lack of recognition of community members, the lack of participatory opportunities, and/or the decimation of community functioning . . . one simply cannot talk of one aspect of justice without it leading to another.
>
> (Schlosberg 2007: 73)

Accordingly, in the case studies discussed in the next four chapters, claims for injustice are not analysed in discrete categories (e.g. procedural justice or justice as recognition); rather, the experience of injustice is explored as interlinked across the various dimensions, with a particular focus on the ways in which they interact to limit or deny broader capabilities. This approach recognises that justice is about 'all that it takes – recognition, participation and more – to be able to fully live the lives we design' (Schlosberg 2007: 34).

Second, in making sense of process claims or 'why things are how they are', it is necessary to acknowledge that environmental inequalities are not explained easily; they are 'produced through multiple interwoven processes' (Walker 2012: 74). In order to understand and address the causes of environmental injustice, 'we necessarily confront relations between the economy, society and the state' (Walker and Bulkeley 2006: 657) across several scales and periods of time (Walker 2012). Careful attention to the ways in which the various actors and institutions interrelate with one another and how power manifests itself in these relationships can provide much insight into the causes of environmental injustice. The law, in particular, is a tool of power which can be wielded in a variety of ways that serves to both enable and constrain actors (Bernal 2011: 145).

The broad literature base consulted in this chapter provides a starting point for exploring the case studies that follow in Chapters 3, 4, 5 and 6. The case studies explore the experiences of various stakeholders in situations of land use conflict over extractive development in rural communities. These experiences are considered in order to locate and explain claims of environmental injustice, highlighting what is fundamentally at stake in disputes over natural resource extraction. Ultimately, this analysis opens up dialogue about alternative approaches to land use decision making, which is considered further in Chapter 7.

3 The Bulga case study

Introduction

In the next four chapters I discuss case studies of coal and gas resource development conflict using the framework of environmental justice. My intent is to present evidence of 'how things are' in terms of perceptions of justice, with a particular focus on the role of the law in each conflict. These explorations discuss the relevant law alongside information about the conflict, including perspectives from those involved in and affected by the disputes. Their 'experiences, viewpoints and the dynamics of power' (Jessup 2013: 84) provide valuable empirical inputs to the analysis.

The Bulga case study, presented in this chapter, concerns conflict over the proposal to extend the integrated Mount Thorley–Warkworth open cut coal mine operation near Bulga, in the Upper Hunter Valley region of NSW, Australia. The case study describes a situation in which, during the phases of project assessment and approval, the state government made changes to the laws regulating the assessment regime to enable the approval of a coal development proposal. The changes arguably disadvantaged the community by prioritising economic benefits at the broader regional and state scales, overriding concerns expressed at the local scale. This left the community feeling disempowered and wronged, and fomented a broader conflict imbued with the language of environmental justice.

I begin the case study by providing an outline of the context in which the conflict arose; that is, the importance of coal to the region around the township of Bulga and, particularly, to the state and Commonwealth governments. I then draw on legal documents (see Table 3.1), development application materials (including departmental assessment reports, and proponent and public submissions) and media reports to describe the project approval process. Throughout the case study, I provide quotes from community members and other stakeholders interviewed (see Table 3.2) to illustrate their perceptions of justice and fairness of the approval process.

Coal mining in the Hunter Valley and NSW

Coal extraction and export commenced in the Hunter Valley in the early 1800s; the initial settlement of the region being for the purposes of mining coal (Evans 2008).

Table 3.1 Legislation, regulations and state policies relevant to the Bulga case study

Relevant laws	Function of the law
Environmental Planning and Assessment Act 1979 (NSW) ('EP&A Act')	Regulates the assessment and approval processes for major developments, such as coal mines, within NSW.
State Environmental Planning Policy (Mining, Petroleum Production and Extractive Industries) 2007 (NSW) ('Mining SEPP')	Provides specific criteria for the assessment of mining proposals, including compatibility with nearby land uses, any impact of transport, the efficiency of resource recovery and mine rehabilitation.
Threatened Species Conservation Act 1995 (NSW)	Aims to conserve threatened species, populations and ecological communities of animals and plants.
Freedom of Information Act 1982 (NSW) ('FOI Act')	Provides a right to request access to documents held by government ministers and agencies in NSW. Access may be denied if the documents are considered to be exempt or if their release is deemed contrary to the public interest.
Government Information Public Access Act 2009 (NSW) ('GIPA Act')	Facilitates public access to government information.

Table 3.2 Key informants for the Bulga case study

Name*	Role
Peter	Resident/community organisation representative
Janet	Resident/community organisation representative
Allan	Resident/community organisation representative
Joanne	Resident/community organisation representative
Bill	Resident/community organisation representative
Carol	Community member
Anne	Community member
Mark	Environmental NGO representative
Monica	Lawyer, environmental advocacy organisation

* All names listed are pseudonyms.

By the early twentieth century, the area had become the biggest coal-producing region in Australia, producing 6.5 million tonnes of coal by 1908 (Cartoscope n.d.). Today, approximately one-quarter of Australia's coal production takes place in the region, particularly in the areas of Muswellbrook, Singleton and Denman (Evans 2008; Cleary 2012).

Up until the 1980s, coal mined throughout the Hunter Valley was predominantly used to fuel domestic power stations. However, in 1981 the Commonwealth government sought to increase coal production and expand foreign investment in the industry to take advantage of growing export opportunities. Following the introduction of substantial subsidies and large infrastructure investments by state and Commonwealth governments, the industry underwent rapid expansion, changing the scope and nature of coal mining in the region (Evans 2008; Cottle and Keys 2014). Dragline open cut mining has largely replaced underground mining (Cottle 2013), with 22 of the region's 31 current mining operations using open cut mining (Climate and Health Alliance 2015). The Hunter Valley coalfields now produce over 100 million tonnes of coal per annum (NSW Government, Department of Industry Resources & Energy 2014), accounting for approximately two-thirds of NSW's coal production (Bureau of Resources and Energy Economics (BREE) 2012). A fifth of this coal is used for domestic electricity generation, with the rest exported (Campbell 2014). The Port of Newcastle, through which coal produced in the Hunter Valley is exported, is the largest coal export port in the world, with the Hunter Valley Coal Chain (consisting of rail and port infrastructure) able to service approximately 1,400 export vessels per annum (NSW Government, Department of Industry Resources & Energy 2016).

The Hunter Valley's contribution to coal production in NSW has been hailed as 'critical to the New South Wales economy' (Cottle and Keys 2014: 3) and the 'bedrock' of the Hunter Valley (Campbell 2014), providing jobs, export earnings and royalty revenue. However, Campbell (2014) argues that such assertions about the importance of coal to the region are rarely contextualised, leading to exaggerations of the economic benefit to Australia and even to the Hunter Valley region. Box 3.1 provides an overview of the contextualised social and economic benefits of the coal industry.

BOX 3.1   The Hunter Valley coal industry in context

Economic contribution: In 2013, the coal industry accounted for only 2–3 per cent of the NSW gross state product, and mining royalties represented less than 2 per cent of total NSW state government revenue (Campbell 2014). While the coal industry accounts for around 22 per cent of NSW's exports, because of foreign ownership, much of the profits from mining go offshore (Campbell 2014). Indeed, only 10 per cent of the saleable coal produced in the Hunter Valley is Australian owned (Campbell 2014). Based on figures from 2010 to 2011, of the then 29 mines in the Hunter Valley and Newcastle Coalfields, only two relatively small mines were majority-Australian owned and 21 were entirely foreign owned (Campbell 2014). This pattern is reflected in coal mine ownership across Australia, where about 83 per cent of the mining sector is foreign owned (Cottle 2013). This trend is likely to continue in the future, with companies throughout Asia (including China, Japan and India), Europe (UK and Switzerland) and North America (Canada and the US) engaged in current or planned coal projects in NSW (Cottle and Keys 2014).

Employment: Over 38,000 people in NSW are employed in the mining sector as at February 2016. This represents 1.01 per cent of the state's full-time/part-time workforce (Australian Bureau of Statistics (ABS) 2016c). In the Hunter Valley region, the percentage figure is higher at about 5 per cent, but the proportion of the Hunter Valley's workforce employed in mining has remaining unchanged since 1976. As Campbell (2014) notes, approximately 95 per cent of the region's workforce is not employed in the coal industry.

Local communities and the environment have borne the brunt of coal mining in the Hunter Valley. The Climate and Health Alliance (2015) reports that, throughout the region, residents have expressed concern with declining air quality, exposure to toxic substances from mine blasting, water contamination, damage to natural ecosystems, aesthetic impacts upon the landscape, social disruption and increased psychological stress. The Hunter Valley is now an air pollution 'hotspot'. Communities in the region report higher rates of hospital admissions for asthma and respiratory disease than the rest of NSW (Climate and Health Alliance 2015). The expansion of coal mines in the region has encroached upon populated areas and agricultural land, and some communities are now 'literally surrounded by coal mines' (Climate and Health Alliance 2015: 7).

Open cut mining, which involves blasting and drilling through rocks to access coal, can operate at a significant scale. At times, the draglines used to excavate the overburden (removed to access the coal seam) operate in an area several kilometres long, leaving behind a large void and a significant amount of dusty spoil. Underground longwall mining also has an impact; notably subsidence, after the coal is removed, which can affect water systems. Estimates are that coal from the Hunter Valley produces approximately 348 million tonnes of carbon dioxide (CO_2) (Campbell 2014), which makes it Australia's largest source of CO_2 emissions (Evans 2008; Climate and Health Alliance 2015). The noise and light pollution from round-the-clock mining operations causes disruption to sleep as well as affecting other quality of life factors (Climate and Health Alliance 2015).

In some areas, mines have been required to purchase residential properties to overcome pollution issues. This can disrupt the social fabric of towns in the process, and some small villages have closed completely. Several studies highlight that the rapid expansion of mining activities throughout the region has had significant negative psychological effects (Higginbotham, Freeman, Connor and Albrecht 2010; Albrecht 2005; Colagiuri, Cochrane and Girgis 2012). Division within communities – over whether to support or oppose mine development, and the distribution of mine impacts – has, in some cases, fractured social cohesion and negatively affected individual and community health (Climate and Health Alliance 2015).

Despite environmental, health and social concerns, exploration activity continues to grow in the Hunter Valley, with further mines planned for the region in the near future (NSW Government, Department of Industry Resources & Energy 2016). It is against this backdrop that the Mount Thorley–Warkworth mine sought to expand its operations near Bulga in the Upper Hunter Valley.

Bulga and the Mount Thorley–Warkworth Mine expansion proposal

Bulga is a small village in the Upper Hunter Valley, NSW, situated approximately 200 kilometres north of Sydney. The village is located approximately 20 kilometres from Singleton, the largest town in the Upper Hunter Valley. In 2011, the population of the Singleton local government area (LGA) was 22,694, with Singleton housing 13,961 people and Bulga 358 (ABS 2011a). The coal mining industry is a large employer in the Singleton area: 24.7 per cent of Singleton's residents are employed in coal mining; and 21.2 per cent of Bulga's residents are employed in the industry (ABS 2016a).

Bulga is situated close to the integrated Mount Thorley–Warkworth open cut coal mine, operated by Coal & Allied (substantially owned by international mining giant, Rio Tinto) (Rio Tinto 2014); the mine is the largest of several coal operations in the Upper Hunter area, employing around 1,300 people. The mine operates around the clock to produce approximately 10 million tonnes of coal per annum (most of which is exported), and has an approved extraction rate of 18 million tonnes of coal per annum, producing thermal and semi-soft coking coal (Rio Tinto n.d.).

Coal has been mined at Mount Thorley–Warkworth complex since 1981, with current operations approved by a development consent granted in 2003 by the Minister for Planning, under the *Environmental Planning and Assessment Act 1979* (NSW) ('*EP&A Act*') (see Table 3.1). The development consent permitted mining through to 2021, and imposed conditions on the operation as set out in a Ministerial Deed of Agreement (the Deed). The Deed provided that certain areas within the leasehold were to be excluded from mining, including:

- The Warkworth Sands Woodland, which is the last known ancient dune landform of its kind, a unique ecological community formed by windblown deposits of sand (Peake, Bell, Tame, Simpson and Curran 2002). It is home to several threatened species, including the squirrel glider, the brown treecreeper, the grey-crowned babbler and the speckled warbler. Around 460 hectares – or 13 per cent – of the original woodland remains.
- Saddleback Ridge, which is a hill between the mine and the village of Bulga that acts as a buffer between Bulga residents and the mine activities.

The Deed required Warkworth Mining Ltd to request that the local Singleton Council establish a conservation zone for these non-disturbance areas. However, no time limit was placed on Warkworth Mining Ltd to make the request and the council was never approached to re-zone the non-disturbance areas (Hannam 2015a).

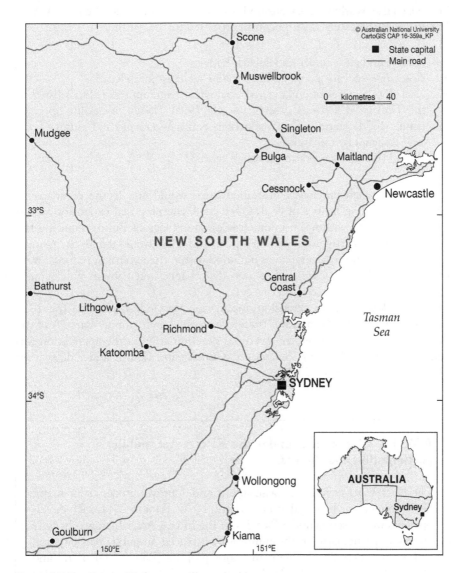

Figure 3.1 Location of Bulga

Warkworth Mining Ltd sought to capitalize on the dramatic increase in the price of coal in 2008, which soared to over US$190 per tonne (IndexMundi 2015), by extending both the term and the physical boundaries of its operations. Several of the areas it wished to mine were previously considered uneconomic because of the high ratio of overburden to be removed per tonne of coal, but the increased price of coal made an extension viable. In 2010, Warkworth Mining

Ltd lodged the 'Warkworth Extension Project' application under the *EP&A Act*.[1] In particular, the extension application proposed to:

- remove and mine under Saddleback Ridge;
- close and excavate a popular local road (Wallaby Scrub Road);
- clear 766 hectares of endangered ecological communities (EECs) listed in the *Threatened Species Conservation Act 1995* (NSW), including approximately 107 hectares of the Warkworth Sands Woodland and Hunter Lowland Redgum Forest; and
- extend the mine's operation through to 2031.

Expansion through previously un-mined areas would enable the company to extract 200 million tonnes of coal. Over time, the proposed expansion would bring the mine to within 2.6 kilometres of the village of Bulga, requiring the acquisition of properties as noise and dust disturbances would render them uninhabitable. In addition, several areas included within the extension proposal were those that had been designated as non-disturbance and habitat management zones.

The Warkworth mine extension application was lodged prior to the 2011 reforms to the *EP&A Act* and was therefore assessed under former Part 3A of the Act (see Box 3.2); however, the eventual determination of the application was conducted by the Planning Assessment Commission (PAC) under delegation from the Minister (see Box 3.3).

BOX 3.2   Assessment under the *EP&A Act*, public participation and the PAC

The *EP&A Act* governs the assessment and approval processes for major developments, such as coal mines, within NSW. Prior to 2011, such developments were assessed under Part 3A of the *EP&A Act*, which was heavily criticised, particularly for the limitations it placed on public participation and environmental assessment in major projects (Bartel, McFarland and Hearfield 2014; Jessup 2013). Introduced in 2005 and further amended in 2008, Part 3A centralised land use decision making by providing the Minister with a broad discretion to 'fast-track' projects deemed to be of state or regional significance, and the ability to delegate decision making to ministerially appointed panels, such as the Planning Assessment Commission (PAC) (Jessup 2013). These amendments reduced opportunities for public participation and 'failed to keep faith with the original reasons for [the Act's] introduction' (Lipman and Stokes 2008: 324). In discussing former Part 3A, one of the legal advisors interviewed for this case study, Monica, remarked:

> I think that was a very dark and hard period for the integrity of the development assessment regime, because it unashamedly just placed massive amounts of discretion in one place of power, namely the Planning Minister. The limitations and guidelines on how that discretion would be exercised was really absent in the regulatory framework.

In 2011, Part 3A was repealed and replaced with a new application and assessment framework, including 'State Significant Developments' under Part 4 (which most coal development applications now fall under). While the process of decision making remained largely consistent with the Part 3A system (including public exhibition and comment processes), the scope for ministerial decision making was reduced; the decision making role of the PAC was extended (see Box 3.3); and changes were made to the appointment process for PAC members to increase transparency and independence of decision making.

BOX 3.3   Decision making role of PAC (from 1 October 2011)

- Act as decision maker for existing Part 3A applications and new State Significant Development applications under Part 4 of the *EP&A Act* where more than 25 objection submissions have been made.
- Review, when requested, development applications and advise on planning and development matters (*EP&A Act* s23D).

In exercising its decision making function, the PAC is required to consider, under s79C of the *EP&A Act*:

- the likely impact of the development (including environmental, social and economic impacts), and
- the requirements of any environmental planning instruments (such as State Environmental Planning Policies (SEPPs), and public submissions. In the case of the Mount Thorley–Warkworth extension application, this was the *Mining SEPP* (see Table 3.1).

Initial assessment of the extension application

The community first became aware of the mine's plans to expand through a local resident who was a member of the Mount Thorley–Warkworth Mine Community Consultative Committee. Community members interviewed for this case study claimed that consultation regarding the application up to that point had been 'non-existent', and there was no prior knowledge of the proponent's intention to expand its operations.

Community members were particularly concerned that the original Deed, which required Warkworth Mining Ltd to request the Singleton Council to re-zone the non-disturbance areas as a conservation zone, was to be contravened by this application. Allan said '[this was] a major breach of faith . . . we felt really cheated . . . we thought that [the Deed] protected us entirely'. When community members requested a copy of the Deed, they were initially denied a copy by the state government on the basis that they were not a party to the Deed and it was confidential. When the community eventually obtained a copy through a Free-dom of Information request (under the *FOI Act*; see Table 3.1), they discovered that the Deed had never been enforced.

Warkworth Mining Ltd's application was placed on exhibition to enable pub-lic comment on 30 April 2010. Under former s 75H(3) of the *EP&A Act*, the minimum period for public exhibition was 30 days. The exhibition period for the mine extension application ran from 30 April until 15 June 2010, providing the public with six weeks to review the application documents and make writ-ten submissions. Many of those interviewed argued that the community was at a disadvantage in this process:

> [T]hey produce reams and reams, thousands of sheets of paper, which most of is technical. They give us six weeks to assess it . . . we've got virtually no show of competing against them in the way of employing consultants to offset it. We can't, six weeks is not enough to run your own tests.
>
> (Peter)

Community members indicated that they struggled to digest, analyse and respond to the complex technical data in the application:

> [It] has been hard I've gotta say – I mean, we've got some really good people here . . . who are perfectly capable of stringing together the right words in the right order . . . [but] you have to be able to pick the key point and address that point.
>
> (Janet)

> We had six weeks to respond. I mean, some of the folk are just very simple country folk and they don't necessarily understand some of the ramifications, or maybe their command of English is not so good.
>
> (Allan)

I hadn't done anything like that since I'd left school, you know, writing essays and doing that sort of [thing]. That was a challenge for me to actually write something and to get across how I wanted it to.

(Joanne)

The tendency for information to be provided in digital form was also a limitation for some people, particularly for older residents who were not computer literate or able to access or read electronic documents. Peter pointed out that some people found it difficult to access additional information about the proposed expansion, partly because of a lack of knowledge of precisely what information they needed and partly because of the belief that access would be denied (based on their experiences in attempting to access the Deed).

Viewed from the standards for procedural fairness discussed in Chapter 2, at even this initial stage of the approval process inequality was evident; that is, information was not provided in a timely manner or in a form that could be easily understood or, in some cases, accessed by the community. These shortcomings constrained the capability of the community to fully participate in the decision making process.

In spite of the difficulties, residents and other concerned stakeholders lodged 102 submissions to the NSW Department of Planning by the end of the exhibition period; 101 of the submissions raised concerns about or were totally opposed to the mine extension. The Department of Planning evaluated the application and proponent's environmental assessment, along with the public submissions regarding the project and Warkworth Mining Ltd's response to the issues raised in the submissions. In October 2011, the Department delivered its assessment report, acknowledging that there would be adverse local environmental and social impacts from the proposed expansion, but concluding that these could be adequately mitigated, managed, offset and/or compensated for. The Department agreed with the mining company that the expansion offered major economic benefits for the region and the state; in doing so, the Department prioritised state and regional economic interests over local interests and the concerns articulated in the submissions.

The project was then referred to the PAC (see Box 3.3) for determination. The PAC considered the Department's recommendations and conducted meetings with Warkworth Mining Ltd, Singleton Council and the department. The PAC also held a public meeting (see Box 3.4) on 10 and 11 November 2011 in Singleton to hear community views about the Department's assessment report and recommendation. At the meeting, individuals were allocated five minutes each to address the PAC, and groups were allocated 15 minutes.

BOX 3.4   Forms of PAC public consultation under the EP&A Act

Under the *EP&A Act*, the PAC can conduct public consultation using public hearings or meetings. **Public hearings** are typically held as a part of

the initial review process for a development, under request from the minister (*EP&A Act* s23D (1)(b)(iii))). **Public meetings** are held as a part of
the decision making process if more than 25 public submissions have been
made that oppose a development.

One of the key distinctions between these two forms of consultation is
the right to access merits review in the Land and Environment Court (see
Box 3.5):

- Where a *public hearing* is held – there are no rights under the *EP&A
  Act* to lodge a merits appeal in the Land and Environment Court
  (s23F).
- Where a *public meeting* is held – merits appeal rights are preserved.

Sixty-nine people spoke at the PAC public meeting about the Bulga expansion.
The issues put to the panel included:

- the failure of the proponent to honour commitments made under the Deed;
- the environmental performance of the mine had already caused significant
  noise and dust disturbance for residents, which would be exacerbated by the
  expansion;
- the considerable threats to EECs and the proposed biodiversity impacts; and
- the impacts upon Aboriginal cultural heritage and the social fabric of the
  Bulga village more generally.

Community members who participated in the PAC public meeting criticised the
forum, noting that the time frame allocated to speakers was particularly 'hope-
less': 'what can you do in five minutes?' Joanne described her discomfort at participating in such a formal meeting ('I don't like to get up and speak in front
of people, I just can't do that – it makes me sick!'), while Mark concluded that
the PAC public meeting did not 'engender public participation', but rather, was
intended to 'marginalise public input'. Others agreed, feeling that their concerns
were simply not 'heard' by the panel members:

> [It felt like] there was a prejudice that, oh well, the mine's put up all this mas-
> sive stuff and they've employed experts and it must be right, and these people
> are just community members, they know nothing.
>
> (Carol)

> The vibe to me was they [the PAC panel members] were just well 'We're here
> just for the sake of being here because we're being told to be here'.
>
> (Joanne)

The literature on procedural fairness (Chapter 2) indicates that once-off 'public hearing style' consultation is an inferior form of public deliberation. The experiences here indicate how the PAC public meeting process fell short of genuine consultation in the eyes of the community. Moreover, these experiences point to issues of recognition. The comments made by Carol and Joanne reiterate that the treatment of individuals and groups in decision making processes are equally as important as the opportunity to participate. In particular, the value placed upon citizen knowledge compared to 'expert' testimony was a critical concern. A failure to recognise this source of information limited the community's capability to influence the processes that determine environmental outcomes.

In early February 2012, the PAC delivered its report, approving the Mount Thorley–Warkworth extension. Despite the previous approval conditions and the Deed, the PAC agreed that the substantial economic benefits offered by proceeding with the project outweighed the actual and potential disadvantages of the expansion. The PAC was satisfied these 'disbenefits' could be adequately managed by a range of conditions, including stricter environmental performance obligations, improved control of dust and noise disturbance through at-source controls (such as noise attenuation equipment) and the provision of new biodiversity offsets. In finding that the current price of coal rendered mine approval 'almost inevitable when the overall benefits of the mines are balanced against the local community impacts', the PAC effectively repositioned the scale of assessment by defining the public interest as overwhelmingly concerned with regional and state economic prosperity, and not with local environmental and social burdens (Jessup 2013). In terms of distributive outcomes, the PAC's decision was perceived by the community to be unjust – particularly given the existence of the Deed, which community members had relied upon in making decisions about their futures (e.g. whether to renovate, or whether to remain in the area). The community considered that the solutions to mitigate and offset the burdens of development were wholly inadequate for the likely harms they would suffer.

Notwithstanding their approval, the PAC did concede that there was 'no doubt' the extension project would have a significant impact upon the Bulga community, potentially threatening the viability of businesses, reducing the value of residential property and changing community identity. In similar cases, they noted, 'communities have either been radically altered in character or become non-viable' (Planning Assessment Commission 2012: 8). The PAC concluded that, in order to alter this trajectory, a clear policy position needed to be developed by government to better guide decision makers as to the weighting of environmental, social and economic impacts. Peter commented that members of the community took these remarks as 'almost a personal invitation to make application for an appeal'.

The community fights back: the appeal to the Land and Environment Court

Following the PAC approval, the Bulga Milbrodale Progress Association (BMPA) lodged a merits appeal (see Box 3.5) in the NSW Land and Environment Court.[2]

The Environmental Defender's Office of NSW (EDO), a community legal centre which specialises in public interest environmental law, represented the BMPA.

BOX 3.5   Merits appeals in the NSW Land and Environment Court

The Land and Environment Court is a specialised environmental court which can hear appeals against administrative decisions on development approvals (see www.lec.justice.nsw.gov.au). Merits appeals by objectors in the Court's Class 1 jurisdiction are only available for State Significant Development (SSD) and designated development under Part 4 of the *EP&A Act*. Only those developments that would have been classed as designated development if they had not become SSD are appellable (*EP&A Act* s98 (4)). Such proceedings require the judge to re-exercise the statutory power of the original decision maker, effectively standing in their place to consider the matter afresh. In its decision making role, the Court considers the environmental, social and economic impacts of the proposed development and the objects of the *EP&A Act,* including the public interest, and principles of ecologically sustainable development (ESD). The Court can also consider new evidence.

   In the Bulga case, a merits appeal was available because the extension application would have been considered designated development but for the operation of Part 3A, and the development determination followed a PAC public meeting rather than a public hearing.

   To lodge an appeal, an appellant is required to have 'objector' status under the *EP&A Act* s98 (4), having lodged a written submission during the public exhibition period. The BMPA had submitted a submission during this period, preserving their right to appeal.

The BMPA raised money for lodging the appeal through various fundraising activities and personal contributions, with the bulk of the funds raised being used for expert opinion evidence. As Allan remarked:

> We have the resources in terms of the people and the arguments, and with the EDO we've got the expert legal people but you need pots and pots of money to be able to put it all through . . . it cost us sixty grand and that's an awful lot of money out of a small community.

Monica, a legal advisor, also noted that the need to have expert evidence was a significant cost for the community:

> I mean we are fortunate that there are some environmental economists that are happy to work on behalf of the community at very reduced rates and

things like that. The difficulty that we will always have . . . is can the community always afford to do a comprehensive analysis and breakdown of all the economic modelling?

Financial and information asymmetry was highlighted in Chapter 2 as a factor that can give rise to procedural injustice. In order for the community to participate in the land use decision making process, they must furnish expert testimony to refute the claims of proponents. This places a heavy burden on communities and constrains their capabilities to participate. The general sense from the interviewees was that the community was the 'underdog', and outnumbered going into the appeal. Bill said: 'it was really a David and Goliath battle' – a metaphor that was invoked frequently in subsequent media coverage of the case (see e.g. Cubby and Rigney 2013).

The hearing ran from August until November in 2012, presided over by Chief Justice Brian Preston. Consistent with their submissions to the Department of Planning and the PAC public meeting, the BMPA argued that the expansion application should have been refused because the development would:

- have significant impacts on biological diversity and EECs;
- increase noise and dust emissions; and
- have serious impacts upon the well-being of the community.

The BMPA tendered evidence of the impact that the mine would have on resident's physical and mental health, given increased noise and dust levels. Further, they argued that the expansion would result in a variety of place transitions in the Bulga community – including the loss of Saddleback Ridge, and the acquisition of several houses – which the BMPA submitted would greatly alter the composition of the community. The BMPA argued that residents were already suffering from stress, sleep deprivation, anxiety and an overall loss of sense of place (known as solastalgia, see Box 3.6), which would be exacerbated by the mine expansion.[3]

BOX 3.6 Solastalgia

A concept developed by Professor Glenn Albrecht, 'solastalgia' is a combination of 'solace' and 'desolation', and reflects the distress and suffering experienced by residents when their place is threatened by significant transformation (Albrecht 2005; Connor, Albrecht, Higginbotham, Freeman and Smith 2004). Albrecht coined the term to describe the feelings of isolation, and the inability to find comfort and solace in one's surroundings as a result of rapid landscape change, which he had researched and witnessed in the Hunter Valley (Albrecht 2005). The term resembles the word *nostalgia*, which is distress caused by an absence from home and implies dislocation from place, but refers to the distress experienced when one remains in their place and experiences the same dislocation:

> [Solastalgia is] . . . the pain experienced when there is recognition that the place where one resides and that one loves is under immediate assault (physical desolation). It is manifest in an attack on one's sense of place, in the erosion of the sense of belonging (identity) to a particular place and a feeling of distress (psychological desolation) about its transformation. It is an intense desire for the place where one is a resident to be maintained in a state that continues to give comfort or solace. Solastalgia is not about looking back to some golden past, nor is it about seeking another place as 'home'. It is the 'lived experience' of the loss of the present as manifest in a feeling of dislocation; of being undermined by forces that destroy the potential for solace to be derived from the present. In short, solastalgia is a form of homesickness one gets when one is still at 'home'.
>
> (Albrecht 2005: 45)

In a historic judgement, on 15 April 2013, three years after the lodgement of the initial project application, Chief Justice Preston upheld BMPA's appeal. He found that, while there were certainly substantial economic benefits to be gained by allowing the extension, these did not outweigh the adverse social and environmental impacts, which included:

- Unacceptable adverse effects of clearing of more than one-quarter of the Warkworth Sands Woodland, threatening it with extinction and negatively impacting fauna habitats and wildlife corridors. Warkworth Mining Ltd had not proposed any avoidance measures; in fact, its plans would nullify the avoidance measures contained in the Deed supporting the existing development consent.
- Insufficient mitigation and offset package to avoid or offset biodiversity losses. Specifically, the offset package did not provide 'like for like', with proposed offset areas of vegetation located a considerable distance from the mine – including one that was more than 100 kilometres away. Chief Justice Preston held that it 'is not appropriate to trade offsets across different ecological communities. Where a project impacts on a specific ecological community, any offset must relate to that same ecological community which is impacted'.[4]
- Unacceptably intrusive noise and dust impacts of the mine extension. Chief Justice Preston found that the costs of the residual noise and dust impacts of the expansion would be borne by the residents of Bulga, yet the benefits would be enjoyed by other parties. Because the proponent would not fully

internalise the external noise and dust costs, His Honour found that the compensation offered to residents was insufficient. In addition, Chief Justice Preston questioned the legality of Warkworth Mining Ltd's proposal to combine both mine sites in setting noise and dust criteria as well as mitigation strategies, as both sites were separately owned and had individual development consents with noise and dust conditions. He also pointed out that the proposed dust and noise mitigation strategies would be difficult to monitor and enforce.[5]

- Unacceptable social impacts. Although Chief Justice Preston limited the reliance that could be placed on the evidence of solastalgia (see Box 3.7), he held that the social impacts of the mine expansion could not be ignored and would 'exacerbate the loss of sense of place, and materially and adversely change the sense of the community, of the residents of Bulga and the surrounding countryside'.[6] His Honour noted that the proponent did not offer sufficient evidence, such as testimony from residents who supported the project, to dispute this impact. Warkworth Mining Ltd's proposed mitigation strategy for the impact of the expansion largely turned on the acquisition of several homes from the village, which Chief Justice Preston held did not deal with the 'subjective or emotional loss occasioned by being dispossessed'.[7]

BOX 3.7   Findings regarding solastalgia in Bulga

Professor Albrecht appeared as a witness for the BMPA and the Court accepted that solastalgia (see Box 3.6) was 'a condition caused by the gradual erosion of the sense of belonging to a particular place and a feeling of distress about its transformation'; and concluded that citizen testimonies revealed 'deep solastalgic distress about the damage that has already been done to their loved landscape and deep anxiety that this level of distress could get even worse as the mine expands towards the edge of the town'. However, the Court found that there were flaws in the research methodology used to collect data in this case. In particular, the sample size of 17 residents was deemed too small and comprised of self-selecting respondents who had objected to the development. Lawyers for Warkworth Mining Ltd also argued that the questionnaires used contained language prompts which could have influenced the responses. Thus, while the Court was open to the concept of solastalgia, the methodology used limited the reliance that could be placed upon the data.

(*Bulga Milbrodale Progress Association Inc v Minister for Planning and Infrastructure and Warkworth Mining Limited* [2013] NSWLEC 48 at [404, 425, 430])

Overall, Chief Justice Preston found that the economic benefits from the project were inadequate to outweigh the costs. Moreover, he found that Warkworth

Mining Ltd's use of an Input–Output Analysis and a Benefit–Cost Analysis to quantify the economic gains of the expansion did not adequately account for issues of distributive justice, particularly failing to fully incorporate the burdens that would be placed on the environment and the people of Bulga.[8] His Honour noted that:

- Warkworth Mining Ltd and its shareholders would benefit from profits;
- the state government would benefit from royalties and taxes;
- the Commonwealth government would receive other income and company taxes;
- the local council would obtain improved local infrastructure; and
- employees and contractors would receive remuneration for work provided.

But:

- the people of Bulga would suffer the burdens of increased noise, dust and other visual and social impacts while receiving little in return;[9]
- the broader community would suffer from the reduced natural and cultural environment in the Bulga area; and
- the components of biological diversity – such as the threatened fauna and EECs – would be significantly disturbed.[10]

In addition, Chief Justice Preston found that intergenerational and intragenerational equity, which are key principles of ESD, had not been adequately considered.[11]

While there had been some financial obstacles for the community in bringing the case to court, members of the community noted that the Land and Environment Court, as an independent forum, enabled a more thorough consideration of all relevant issues compared with the PAC public meeting: '[W]e were made to feel that we actually counted as a community. Someone actually heard what we were saying' (Joanne).

Interestingly, Chief Justice Preston's findings of fact about the impacts of the extension were not distant from those that were previously reached by the PAC. However, his weighting of all relevant issues reoriented the scale of public interest assessment to one which more clearly linked the broader benefits of consumption to the intense environmental and social burdens that would be borne by the community (Jessup 2013; Bickerstaff and Agyeman 2009). This was, arguably, consistent with the provisions of the overarching legislation (particularly s 79C of the *EP&A Act* (see Box 3.3 earlier); Dwyer 2014).

An attempt to regain power: the Supreme Court appeal

On 22 April 2013, days after Chief Justice Preston's decision, Warkworth lodged an appeal to the Court of Appeal, the appellate jurisdiction of the Supreme Court of NSW (the highest court in the state). Representatives of Rio Tinto, the parent company of the mine operator, justified the appeal on the basis that the Land and Environment Court had 'disagreed with the outcome of a rigorous planning process that had determined the project was in the overall public interest' (Tasker 2013). The grounds of the appeal included, among other issues, that:

- Chief Justice Preston had erred in law by failing to give appropriate weight to the Director-General of Planning's assessment report from 2011;
- His Honour had failed to address the significant economic benefits and positive impacts of the project; and
- Warkworth Mining Ltd had been denied procedural fairness regarding several matters, including the weight applied to expert testimony regarding noise impacts.

Warkworth also announced in the media that 40 employees would be made redundant, claiming that the rejection of the mine's approval threatened the ongoing viability of the mine (Coal & Allied 2013). Warkworth Mining Ltd further used the issue of job vulnerability in an application to the Supreme Court to expedite proceedings (Wen 2013), which the Supreme Court approved. These actions effectively entrenched the conflict over the mine expansion as a struggle over scale (see Chapter 2).

The Minister for Planning then lodged a cross appeal on the basis that the Bulga case decision had potential implications on the assessment of other mining projects (Battersby 2013), particularly the weighting applied to the initial assessment offered by the Director-General of Planning. The cross appeal was a further example of a scale frame, which was 'stacked' (Sica 2015) on top of what had already been advanced by the proponent to provide further legitimacy for the broader economic scale. Members of the community were critical of the Minister joining the appeal:

> It's bad enough trying to fight a large company let alone trying to fight your own government . . . your own government . . . shouldn't be holding hands with a multi-billion dollar company. They should be there protecting the community and that didn't happen.
>
> (Joanne)

> Supporting [the appeal] just proved whose side the Department's on. Why else would they be there if it wasn't to get the mine extension through?
>
> (Bill)

The Supreme Court appeal hearing commenced in August 2013. In the meantime, the 'higher' regional/state scale was given further support by other discursive tactics, while the local scale was downplayed through 'counter-scale frames' (Kurtz 2003). Rio Tinto's London-based Chief Executive of Energy, Harry Kenyon-Slaney (as a representative of Warkworth's parent company) condemned Chief Justice Preston's decision in national newspaper opinion pieces, and argued that courts were not appropriate venues for determining major mining projects of state significance (Kenyon-Slaney 2013a, 2013b). Members of the community noted that the proponent had far greater access to the media than they did:

> [W]e occasionally get letters published, we occasionally get articles published, but the mine puts in full page newspaper ads every week. I mean, come on, that's big money for the newspaper. They can't possibly refuse that.
>
> (Carol)

> [A]nother thing they did . . . Coal & Allied sponsored the Newcastle Knights football team [a team in the national rugby league competition] for a number of years . . . to call in a favour they had the team all dressed up in Coal & Allied orange shirts and paraded them around and they even dedicated one football game to Mount Thorley–Warkworth . . . the football team would never have seen the mine, know nothing about it but they're being asked a favour because they're sponsored by Coal & Allied. That publicity . . . we can't match that sort of thing.
>
> (Bill)

The impact of this publicity, as Bill said, was that, to the 'average citizen', the economic benefits of mining (and by extension, the approval of the Warkworth development) became increasingly viewed as more important than the impacts upon the community and the environment. With little contextual information provided in claims about the benefits of mining (Campbell 2014; see also Box 3.1 earlier), the role and the necessity of the coal industry, and the Warkworth extension in particular, became greatly inflated in public discourse. As Allan noted, the BMPA was portrayed by the proponent as a small group of 'troublemakers'. He maintained that those who objected to the extension were not a minority of self-interested 'dissidents': 'There were 200-odd [people] at a community meeting . . . that kind of denial that there is community support [for the BMPA] . . . its crook.'

   In addition to targeting the media to downplay local concerns, representatives of Warkworth's parent company (Rio Tinto) and the mine operator (Coal & Allied) lobbied the NSW government (including the Treasurer, the Minister for Planning, the Minister for Energy and Resources and the Minister for Trade and Investment) (Hannam 2014a). A private meeting between Harry Kenyon-Slaney and the then premier, Barry O'Farrell, took place (Lagan 2013). A request at the time for information about the contents of the meeting under FOI (see Table 3.1) were unsuccessful (Lagan 2013). The community's requests for meetings with the Premier were also denied:

> When [Kenyon-Slaney] flew from England and had an instant meeting with the Premier, we wrote to the Premier and said 'Look you must be aware there are two sides to this and . . . we'd like to meet with you'. Now it took us two months to get a response and the response said just basically 'No', he wasn't going to meet with us. Too busy doing other things.
>
> (Peter)

Members of the community consistently cited the ability of the industry to access politicians as unfair and evidence of a lack of recognition of the local community. Interviewees repeatedly spoke of being disillusioned; Peter remarked:

> There is too much access by lobbyists and developers . . . to the government, which brings bearing on their decisions . . . The same avenues are not open to the general public, or communities such as ours. So it's an unfair process.

In a further effort to reinforce the broader economic scale, Rio Tinto submitted legal advice to the NSW Premier, which claimed that the Land and Environment Court judgement 'sets a precedent with the potential to threaten all major project approvals and applications . . . and therefore also threatens the broader economic development of the State'.[12] They strongly urged the government to consider legislating to validate the PAC's original determination in order to 'rectify issues arising from Court decisions which do not accord with Government policy'. The government subsequently acted on this advice.

Reframing scales, re-establishing power

In July 2013, weeks before the Court of Appeal hearing commenced, the NSW government proposed amendments to the *Mining SEPP* (see Table 3.1 and Box 3.8). Of particular relevance to the Bulga case was the insertion of a new factor for consideration by decision makers: that they must prioritise the significance of a mineral resource as their 'principal consideration' with regard to the economic benefit of developing the resource at both regional and state levels (*Mining SEPP* Reg 12AA).

BOX 3.8   SEPPs

State Environmental Planning Policies (SEPPs) are environmental planning instruments provided under the *EP&A Act*, which the Governor makes on recommendation from the Minister (s37). If the Minister thinks it is necessary, they can seek public submissions regarding a SEPP under s38 of the *EP&A Act*. The *Mining SEPP* provides guidance to decision makers for the assessment and approval of mine development applications under the *EP&A Act*, listing factors that the decision maker should take into consideration (under Part 3). This includes the compatibility of the mine with surrounding land uses, natural resource management, the efficiency of recovering the mineral resource, the impacts of transport to and from the site on public roads and residential areas, and post-mine land rehabilitation.

Following a two-week public exhibition period, the amendments to the *Mining SEPP* were adopted in November 2013. The final amendment added examples of economic benefits to guide decision makers when considering the significance of a resource, including the level of royalties generated, the amount of jobs created, and increased regional expenditure and capital investment. Overall, the amended *Mining SEPP* required decision makers (whether the minister, the PAC or a judge) to prioritise the economic benefits to be generated by a proposal when determining whether to approve it.

While the overarching legislation, the *EP&A Act*, still provided that discretionary standards, such as environmental and social impacts and the public interest, be taken into account in assessment (s79C), the *Mining SEPP* amendment introduced a specific weighting of economic factors, prioritising the regional and state scale. This reinforced the scale frames already advanced in favour of development, which would subsequently prove useful to the proponent. Critics argued that the amendment distorted the planning process and left little doubt as to the government's intentions (EDO 2013), but the then-Minister for Resources and Energy, Chris Hartcher, claimed that the timing of the *Mining SEPP* amendment was merely coincidental (Thompson 2013). Members of the community, meanwhile, saw it as a direct 'retaliation' and 'response' to their victory in the Land and Environment Court. During interviews for this case study, the *Mining SEPP* amendments were described by various members of the community as an attempt to 'manipulate outcomes', an 'act of bastardry' and even 'unconstitutional':[13]

> Clearly the government has the right to change the rules, I mean they've got to refine their rules . . . but this one was specifically aimed at overturning a properly considered judgement and not only overturning it, but to actually change the rules such that no one will ever be able to win again because of the way they've weighted this.
>
> (Peter)

> I think it was a nasty blow to the community in the sense that the community engaged in such good faith . . . to kind of throw that in, I think it was fairly clear it was to try to respond to [the Land and Environment Court judgement] and placate companies like Rio Tinto and placate the Minerals Council.
>
> (Monica)

As it eventuated, before the appeal had been decided, Warkworth Mining Ltd sought to test the new provisions of the *SEPP* by lodging an application to modify the 2003 approval (known as 'Modification Six').[14] The application sought to extend the 2003 approval by 30 hectares to enable the extraction of a further 13 million tonnes of coal.

On 4 December 2013, Modification Six was sent to the PAC for determination. Although such applications are not required to be placed on public exhibition it was decided to exhibit it for a two-week period. Members of the community complained they were once again taken by surprise:

> When they hit us with [Modification Six], we heard about it on the radio at lunchtime and at three o'clock in the afternoon the manager rang [consultative committee members] to announce what their plans were . . . We'd had a [consultative committee] meeting only a few weeks before that and there was no mention whatsoever, so they kept the community in the dark.
>
> (Bill)

Once again, the pragmatics of participation were such that community members felt constrained in their ability to engage effectively in the decision making process:

> It was all done with indecent haste. We were given the minimal time to object. Because of when it was [December], a lot of people were on vacation . . . so we couldn't get hold of them for expert opinions.
>
> (Allan)

In September 2013, after the completion of the Supreme Court hearing but with judgement still outstanding, the BMPA became aware that the Deed had been amended to allow development in the non-disturbance areas (Hannam 2015a). Members of the community were angry that the Deed had been quietly changed and that they had no recourse to challenge it: '[Changing the Deed] may be legal, but people here bought and sold, didn't do renovations or did renovations, they made certain plans based on the fact that it was in place' (Peter).

In subsequently justifying the changes, Rio Tinto claimed that the original Deed was an early attempt by Warkworth Mining Ltd and the government at offsetting and was not consistent with the new *Biodiversity Offsets Policy for Major Projects* (Hannam 2015a) which broadened the ways in which biodiversity offsetting may be managed (see Box 3.9). In effect, by anchoring the alteration of the Deed to 'current policy' imperatives, another scale frame was strategically deployed in favour of development.

In December 2013, while the Modification Six application process was underway, the community suffered a further blow to its capacity to advance its interests. The Commonwealth Attorney-General's Department announced that it would be removing approximately A$10 million of funding from the nine EDOs throughout Australia, including the NSW EDO which acted for the BMPA in their original case and the subsequent appeal. This followed the removal of Legal Aid funding for public interest environmental litigation by the NSW state government, and the prohibition on the use of public funds to provide legal advice to lobby groups (Aston 2013; Cubby 2013). Documents obtained under the *FOI Act* revealed that the Minerals Council and the Australian Coal Association had lobbied the NSW government to withdraw funding from the EDO (Aston 2013). Reflecting on these government actions, Monica commented:

> I think it was extremely short-sighted, un-strategic, and absent of an engagement in good tactics because at the end of the day, and it can be summarised in that cliché of you will lose ultimately if you play the man and not the ball . . . I think ultimately it backfired on those that called for [the EDO] to disappear because . . . it made the community really more engaged to stand up more defiantly.

At the end of January 2014, still before the Supreme Court had reached its decision on the extension application appeal, the PAC approved the Modification Six application. While the community raised concerns with the PAC regarding the approval of Modification Six during the ongoing Supreme Court appeal process, the PAC found that the modification was a separate matter from the larger extension application (PAC 2014c); it was deemed small in size, considered unlikely to impact Saddleback Ridge or the Warkworth Sands EEC, and to have relatively marginal impacts in terms of noise, dust and biodiversity loss because the 'small amount' of coal that would be extracted was readily accessible. Because these impacts were considered minor, the PAC did not seek to verify the proponent's claims regarding the threats to employment, and other economic costs and benefits. Overall, the PAC considered that Modification Six would provide 'greater certainty for the mine's existing workforce' and was a 'logical extension of the existing mine' (PAC 2014c: 2). According to members of the community, however, the approval of Modification Six was a 'conspiracy' between the department and the proponent, and, in Peter's words, '[it was the] start of the first stage of a bigger project'.

Consolidating the power imbalance

Changes to the *Mining SEPP* and the approval of Modification Six set the scene for a fresh development application. On 1 April 2014, Warkworth Mining Ltd lodged requests with the Minister for Planning for Director-General's Requirements (DGRs) for two new projects: the Warkworth Continuation Project, and the Mount Thorley Continuation Project.[15] The company lodged two separate applications, taking into account Chief Justice Preston's criticism of the integrated Warkworth mine extension application. While the expansion footprint of the operation remained the same, the new applications contained modifications in light of the Land and Environment Court's decision. In particular, Warkworth Mining Ltd revised their biodiversity offset strategy and their social and economic assessments, and proposed other operational changes along with additional cultural heritage and infrastructure commitments.

On 7 April 2014, almost one year after Chief Justice Preston's decision to overturn approval of the expansion,[16] and a matter of days after the two continuation applications were lodged, the Supreme Court handed down its judgement in the appeal. They affirmed Chief Justice Preston's decision, rejecting all grounds of appeal raised by Warkworth Mining Ltd (as the appellant). In particular, the Supreme Court of Appeal:

- Found that Chief Justice Preston's rejection of expert testimony offered by the proponent on acoustic impacts (which was unchallenged by the BMPA) without expressing his intention to discount this evidence was not considered a failure to afford procedural fairness to the proponent.[17]
- Held that Chief Justice Preston had not erred at law in his consideration of the public interest by failing to account for the positive economic impacts of the project, the statutory provision regarding the public interest is general, and the need to balance principles of ESD with economic considerations.[18]
- Dismissed all other grounds of appeal advanced by Warkworth, including claims that Chief Justice Preston had failed to consider the implications of new mitigation measures proposed during the hearing.

The refusal of the extension application meant that the mine was operating under the existing development consent granted in 2003, which permitted mining through to 2021.

However, this apparent victory for the community was to have little impact. The lodgement of the two new extension applications under the changed *Mining SEPP* entitled Warkworth Mining Ltd to again seek approval for their expansion project. Several interviewees expressed outrage that the proponent had been able to submit a substantially similar application to the one that the Land and Environment Court and the Supreme Court had previously rejected:

> It's emblematic of how unfair the whole system is in New South Wales, where you've got a community that has proven now twice, in courts of law, that the mine should not go ahead because the costs outweigh the benefits, and the government's response is to change the law . . . and the company's response is to immediately reapply for the same coal mine.
>
> (Mark)

> The courts have said twice you can't do it – how do they think, why do they think, that this is acceptable now? It wasn't then!
>
> (Janet)

> To say that we were shocked is really an understatement. Gob-smacked doesn't even come near it. I mean, how could they be so arrogant as to just do the whole thing again? It shows a contempt of the court process and a total disdain for the community . . . They've had the hide to say this is a different application. It's not. It's the same footprint . . . there are different bits around the periphery, yes, to do with dumping of spoil in Mount Thorley mine, but it's the same application.
>
> (Allan)

I mean that project was determined and it was finally determined, so to tweak the law, tweak around the edges, and have another go . . . how does a regulatory system really genuinely expect the community to be able to engage in good faith in a system that does that? To see essentially the same project come before a decision maker again . . . [it] just reeks.

(Monica)

The DGRs were issued on 22 May 2014, to which the proponent responded with a lengthy environmental impact report a little over three weeks later. Media reports alleged that the swift response was facilitated by a process of 'workshopping' with the Department of Planning, amounting to 'regulatory capture of the worst kind' (Hannam 2014b). Interviewees also expressed concern that the proponent and the Department had worked 'in very close concert... [it's a] closed door approach and the transparency and accountability of the Department of Planning is a major worry' (Monica). As Peter put it, there was a perception in the community that there was 'a lot of work being done in the background . . . [with] no public participation'. Bill concluded that the community felt like a marginal stakeholder in the process, and that Warkworth Mining Ltd and the Department had treated them 'like the stuff they dig out – dirt'.

The two Continuation Project applications were on public exhibition between 25 June and 6 August. Members of the community again expressed frustration with their inability to engage effectively in the submission process:

We had six weeks but it was probably a week or more before we got our hands on [the project documents] and there's just a hell of a lot of reading, there's 4,000 pages in all and not a lot of time to go through it.

(Bill)

Out of the 1,926 submissions received, 1,638 supported the project. Submissions in support were received largely from those working in the mining industry or associated businesses. In November, the Department recommended, in its environmental assessment report, that both the Mount Thorley and Warkworth Continuation Projects should be approved. In their report on the Warkworth Continuation Project, the Department reiterated that, despite the fact that the original extension application had been denied by two courts, Warkworth Mining Ltd was legally entitled to lodge a new application for a modified project and the Department was obliged to assess the application on its merits (Department of Planning and Environment 2014b).

The Department noted, in its report, that while the amended *Mining SEPP* had prioritised the significance of the resource as a factor for assessment, environmental and social impacts were still important concerns under other environmental planning instruments. Moreover, s 79C of the overarching *EP&A Act* requires factors other than environmental planning instruments to be weighed up by the decision maker, including the impacts of development, the suitability of development, any public submissions made regarding the development and the public interest.

Nonetheless, it was clear that the changed *Mining SEPP* was central to the Department's assessment, with the project's potentially 'significant economic benefits to the Singleton LGA, the Hunter region and to the State' given greater weight than the amenity and biodiversity impacts of the project on the Bulga community. Members of the community felt that their submissions, in particular their expert report on acoustic disturbance, had been overlooked by the Department in their assessment:

> I heard all the rumours about the unfairness of the process. I was surprised, though, when the Department of Planning came down with the recommendation to approve it, and I thought . . . after we'd done all these submissions . . . we had three, four consultants' reports backing that up, and I realised that maybe this is not all above board because the Department of Planning never referred to any of our consultants' reports at all [in the assessment report].

> (Peter)

'They've moved the goalposts': assessment of the Continuation applications

On 6 November 2014, the Minister requested that the PAC conduct a review of the merits of the Continuation Projects and conduct a public hearing (which would preclude the ability to seek merits review of the eventual PAC determination in the Land and Environment Court: *EP&A Act* s 23F – see Box 3.3). The community felt that the ministerial direction to hold a PAC hearing was a strategic measure to prevent them from again appealing to the Land and Environment Court if the PAC eventually approved the Continuation Project applications:

> Essentially what that means is that we've lost our opportunity to have a judge take an unbiased look at a project and make a decision on its merits. It's been replaced by this dodgy rubber stamping operation where three former bureaucrats sit at a table and give people five minutes to say a few words to them, all the while they sit there and look at their watches waiting for it to all be over, and then go and have closed meetings with the proponent themselves and with the Planning Department who are also pushing for it to be approved. It's a completely one-sided process that I feel is designed to disempower communities rather than to give them the opportunity to actually have input into a process.

> (Mark)

Community members were also concerned by the overwhelming number of mine employees in attendance at the public hearing. As Bill noted, this created an uncomfortable environment for delivering their statements, inhibiting participation: '[the employees] filled up the room, it wasn't a very big room . . . it was intimidatory'.

The literature on procedural justice, considered in Chapter 2, notes the importance of access to independent and impartial review functions. The distinction between a merits review in the Land and Environment Court and a PAC public hearing is vast, as EDO Principal Solicitor, Sue Higginson, has observed:

> A public hearing before the PAC and a merits hearing before the Land and Environment Court cannot be compared. A PAC public hearing usually takes one day with members of the community allowed about 15 to 20 minutes to tell the commission why they support or object to the project. On the other hand, a hearing on the merits in the Land and Environment Court can take a fresh and objective look at a major development proposal, including hearing from expert witnesses whose opinions are tested and challenged by the parties.
>
> (Higginson 2014)

In its previous report into 'Anti-Corruption Safeguards and the NSW Planning System', the NSW Independent Commission Against Corruption (ICAC) was critical of the limited availability of third party appeal rights under the *EP&A Act*, noting that without such appeals 'an important check on executive government is absent' which may create 'an opportunity for corrupt conduct to occur' (ICAC 2012: 22). The ICAC had recommended that the NSW government consider expanding access to third party merit appeals, particularly for cases where the proposed development is significant and controversial.

While the PAC review (and associated public hearing) was underway, the NSW government introduced two new policies:

1. the *Biodiversity Offsets Policy for Major Projects* (in October 2014) (see Box 3.9), and
2. the *Voluntary Land Acquisition and Management Policy* (in December 2014) (see Box 3.10).

These two instruments would lend further support to the proponent's applications by providing a clear policy platform against which the impacts of the projects could be measured and justified.

BOX 3.9   The *Biodiversity Offsets Policy for Major Projects*

The difficulty in assessing the impacts on biodiversity from a development and then determining equivalent offsets arose in the original Land and Environment Court appeal and featured heavily in Chief Justice Preston's judgement. The *Biodiversity Offsets Policy* attempts to deal with the conundrum of offsetting for major development projects by providing that

---

biodiversity liability can be addressed by direct offsets *or* by translating liability into a monetary value. In particular, the policy introduced new variation rules if like-for-like offsets cannot be identified, allowing vegetation, flora and fauna to be offset from a much broader range of ecological categories (for example, vegetation may be offset with vegetation from the same formation).

The policy also permits the use of supplementary measures in lieu of offsets, allowing proponents to provide funds to improve biodiversity values (such as contributing to a threatened species recovery program) where appropriate offset sites cannot be found. It also permits mine site rehabilitation to be counted as an offset. The EDO has expressed concern that the use of such indirect offsets 'essentially allows a developer to buy their way out of a difficult offsetting requirement', and constitutes a fundamental breach of the like-for-like principle (EDO 2014: 9).

BOX 3.10 The *Voluntary Land Acquisition and Management Policy*

This policy sets out the noise and dust criteria which trigger mitigation and acquisition rights for State Significant mining, petroleum and extractive industry developments. The policy characterises noise and dust impacts from exceedences of assessment criteria (contained in other policy instruments, such as the NSW Industrial Noise Policy), which trigger voluntary mitigation and acquisition rights implemented as approval conditions by decision makers. The policy is listed as a matter for consideration under Reg 12A of the *Mining SEPP*.

Media reports allege that industry lobbying shaped the development of this policy (Davies 2016a, 2016b). The NSW Premier's meeting disclosure summary for that period shows that a meeting took place between the Premier, the Minister for Resources and Energy, the NSW Minerals Council and several coal companies (including Rio Tinto and Watermark Coal) on November 7 'to discuss government policies relating to the mining industry' (Lock The Gate 2016a); and documents obtained under the *GIPA Act* by Lock The Gate revealed that changes were made to the draft policy shortly after this meeting took place (Davies 2016a, 2016b; Lock The Gate 2016a). The changes limited the criteria for mitigation and acquisition from what had been originally proposed in the draft policy, effectively reducing the number of properties that proponents would be required to offer mitigation or acquisition rights to as a condition of approval.

A significant issue for residents, and a critical issue in the previous Land and Environment Court decision, was noise impacts from mine operations and their satisfactory measurement. A reliable assessment of background noise levels is important, because project-specific noise levels (exceedences of which may provide access to mitigation and acquisition rights) are measured against background noise levels. The PAC was satisfied with the proponent's measurement of background noise submitted in its Environmental Impact Statement (EIS) documents and found that the Department's assessment of predicted noise exceedences for the project was consistent with current policy (in particular, the new *Voluntary Land Acquisition Management Policy*) (PAC 2015a: 28–37). With the bulk of the likely noise exceedences now considered 'negligible' under the new policy, mitigation and acquisition rights would be extended to fewer properties than what was initially proposed under the original Warkworth extension application approval in 2010.[19] The PAC acknowledged the adverse social impacts of proceeding with the proposal, including the impact upon property values and residential amenity, but concluded that these could be compensated for. In particular, they encouraged the consideration of options, including compensating property owners who wished to sell, developing a public works enhancement strategy and even relocating the village of Bulga at the expense of the government and the proponent (PAC 2015a, 2015b).

In May 2015, the Department referred the application back to the PAC for final determination. The PAC held a public meeting in June as a part of the determination process, but before they could make their final decision, the recently appointed Planning Minister, Rob Stokes, announced that the *Mining SEPP* was again to be amended. Specifically, the amendment repealed Clause 12AA, the controversial 'resource significance' requirement, following 'community and stakeholder concern that the social and environmental impacts of a proposal are not being adequately considered' (Department of Planning and Environment 2015). Minister Stokes requested a second PAC review in light of these changes.

The PAC commenced their second review process for the two Continuation Projects in September 2015, including a further public hearing. They delivered their second review reports in October 2015, again finding that the project was approvable (PAC 2015c, 2015d). The BMPA, and other objectors, had submitted to the PAC that, in light of the most recent *Mining SEPP* amendment, the assessment of the Continuation applications should revert to the Land and Environment Court's previous refusal. The PAC, however, disagreed; they did not believe that Clause 12AA was the determining factor in their first review recommendation that the Continuation Projects were approvable, and considered that they had appropriately weighed the economic, social and environmental impacts of the projects (PAC 2015c: 5, 2015d: 5). The PAC believed that the latest Warkworth and Mount Thorley Continuation proposals sufficiently dealt with issues raised in the Land and Environment Court's judgement and adhered to the recent policy changes (namely, the *Biodiversity Offsets Policy* and the *Voluntary Land Acquisition and Management Policy*).

In November 2015, the PAC handed down their final decision that the two Continuation Projects should be approved. The PAC reiterated that they had considered all of the economic, social and environmental impacts (PAC 2015e, 2015f). They repeated that there had been several recent changes to government policy and legislation of relevance to the assessment of both applications, and that they were satisfied that the Continuation Projects were consistent with these changes. The need to be consistent with the changed policy environment enabled the PAC to defend itself against accusations of inconsistency with the previous Land and Environment Court judgement; in this case, the new policies on biodiversity offsets and voluntary land management were invoked to reject any suggestions of outcome unfairness. The PAC concluded that the final conditions of consent proposed (which included additional acquisition rights,[20] stricter monitoring of noise and dust disturbances, and ongoing management of biodiversity regeneration activities) would be appropriate to manage the environmental and social impacts of the projects.

While the Minister's direction to the PAC in November 2014 to hold a public hearing closed off merits appeal rights, the BMPA was still able to seek review on grounds of legal error. In early 2016, the BMPA lodged a judicial review appeal against the final decision of the PAC in the Land and Environment Court, arguing that the PAC had erred in its decision making function. However, in May 2016, the BMPA discontinued the appeal, citing the high threshold required to demonstrate an error of law as a significant barrier to their claim (BMPA 2016).

By mid-2016, the expansion project had commenced and, having exhausted all legal avenues, members of the community engaged in a peaceful protest to object to the expansion approval (Nichols 2016). The protest resulted in the arrest of two Indigenous elders who refused to move on following police orders (Nichols 2016). A further protest took place in Sydney in August, with over 200 people gathering to establish the symbolic 'New Bulga' behind Parliament House and call for greater public participation and merits appeal rights in the land use planning system (Nguyen 2016).

Conclusion

National advocacy group Lock The Gate recently remarked that the Bulga case:

> [I]s an icon for everything that is wrong with the way NSW deals with coal mines. The system is unjust and perverse. It is breaking Hunter communities, who are never now allowed the right to challenge the merits of mining approvals in court. Bad decisions are being made and there is no recourse to justice.
>
> (Lock The Gate 2016b)

When viewed from the perspective of environmental justice, the Bulga case study provides an example of how environmental injustice arose through inequitable distribution, ineffective public participation mechanisms, misrecognition and, ultimately, constraints placed upon the capabilities of the community to function in their desired manner.

The distributive outcomes of the PAC's final decision were considered by the community to be grossly unjust; the impact would be disproportionately felt by local residents, and the measures proposed to mitigate, offset and compensate for losses were considered inadequate given the nature of the potential harms to be suffered. This was a clear finding of Chief Justice Preston in the original Land and Environment Court decision; but the multiple policy changes implemented following his decision provided the means for ignoring the community's concerns and facilitating the unequal distribution of harms and benefits. The fact that the Deed, which had provided that certain areas would be protected in perpetuity, could be changed and not upheld by the government and the proponent, further exacerbated the community's sense of injustice.

In terms of procedural justice, community members interviewed for this case study considered opportunities for public participation to be 'torturous' and imposing significant burdens on citizens who wished to exercise their rights to participate. The burdens included insufficient time frames provided to the community to enable careful consideration of the relevant information – particularly when the materials provided by proponents were lengthy and/or overly technical – to prepare submissions; as well as the costs involved with obtaining expert advice to evaluate the proponent's proposals. The Bulga case illustrates that, despite the existence of participatory opportunities, a variety of barriers can limit the capabilities of communities to participate effectively in environmental decision making.

Interviewees complained that the participatory forums, that is, the PAC public hearings and public meetings, offered little opportunity for genuine consultation and dialogue. The limited time allocated to community members to speak and the formal nature of proceedings left little room for authentic deliberation. As Allan noted:

> [They could] come back to the objectors and say 'well, they say this, what do you say?' Okay, so it will take longer but maybe it needs to take longer to make sure that the process is not just vigorous but also fair to all parties.

Tied to these concerns were perceptions of misrecognition, in which, as Mark said: 'people spend all their time writing huge submissions, responding to reports and then [get] ignored'. Similar to Conrad et al.'s (2011) findings, interviewees in the Bulga case believed that government decision makers were 'hearing but not listening' to their concerns, and felt that their comments were taken note of but were not seen to be acted upon:

> [The government appears] to take the community's wishes on board and to incorporate the will of the local community into these processes but it's a sham, they don't actually listen to anything that the communities say . . . the whole thing is predetermined from the outset.
>
> (Mark)

The need for the public to source 'expert' testimony in order to effectively participate in decision making processes, and the limited value placed upon citizen knowledge as a source of evidence, further impaired recognition and constrained the community's ability to influence outcomes. For members of the community, these place-based concerns were fundamental; they went to the very heart of their opposition to the extension:

> We don't want to see Bulga die for a whole number of reasons . . . in six years' time we're going to have our 200th birthday and there's not many places in Australia, particularly not little villages, that can say they've been around 200 years . . . it's pretty important to us – this is the site of the first entry to the Upper Hunter.
>
> (Allan)

> We are just not going to write a submission and that's it . . . we care about our community . . . my family's been in this area for over 100 years. I'd like my son to be here as another generation but if the project goes ahead that's not going to happen.
>
> (Joanne)

Monica believed that there was already scope within the present framework to account for such place-based impacts:

> We're really not identifying to the extent we should, given the expertise out there, what the impacts are. I mean we had Glenn Albrecht talking about solastalgia and it just displayed how much scope there really is out there that the regulatory system is just not engaging with in terms of exploring impacts. I actually think the current laws provide ample space for the integration of this stuff; if we were looking at it to its fullest, in the absolute spirit of what was enacted, the words don't hold us back.

However, as many interviewees noted, the political influence of the proponent and the mining sector more generally ensured that place-based concerns would receive limited recognition. In particular, interviewees viewed the proponent's ability to leverage the government to reframe the scale of assessment following the Land and Environment Court decision as an example of the overwhelming power of the industry and the community's relative powerlessness. On the whole, members of the community felt like the government and the proponent had consistently 'changed the rules' and 'moved the goalposts', ensuring that the community were constantly outmanoeuvred: 'I think the worst thing is that you approach each battle knowing that the odds are stacked against you' (Allan). As Mark summarised:

> I guess the feeling out there on the ground is that the regime . . . purports to be independent, but the way that it's actually carried out, people feel very disenfranchised by the process . . . and feel like that there are

these complex planning arrangements in place that, at the end of the day, amount to rubber stamping. It's anything but a fair process that encourages communities to engage . . . rather, it's something that wears communities down and sends a very strong message to them that there's absolutely no point in fighting against a coal mine in your area because you're not going to win.

Some interviewees even went as far as to characterise the actions of the government as corrupt:

The fact that the government can now change the parameters within which decisions are made and also remove some of our rights under that, is not right . . . it is not against the law, but the way the government is acting in corrupting the law, so that it favours a particular section of industry for the purposes of profit, is not right.

(Peter)

[Some think that] handing someone some money in a brown paper bag, that's corruption, but corruption can come out in any form . . . unfairly favouring one party against another, that is, in my opinion, corruption.

(Bill)

Through various scale framing processes – achieved particularly through their access to the media, but also their formidable lobbying efforts – both the proponent and the wider industry were able to position the extension application as urgent and necessary for the economic growth of the region and the state. A broad and historical narrative that the Hunter Valley coal industry was 'vital' to the economy provided further support to these scale-framing efforts (Campbell 2014). The simultaneous downplaying and discrediting of the community's concerns led those opposed to development to be characterised as self-interested, further marginalising their interests. The community had some opportunities to reach a wider audience with their story, but they did not enjoy the same level of exposure as the proponent and the industry.

The government deployed extra-legal acts to endorse the broader regional/ state economic scale and their subsequent introduction of new 'institutional infrastructure' normalised it (Chung and Xu 2015: 2; Jessup 2013). Specifically, the implementation of the *Mining SEPP* prioritised the significance of mineral resources in terms of economic benefit at regional and state levels as a criteria for assessment; the *Biodiversity Offsets Policy for Major Projects* expanded the scales at which offsetting activities could be counted towards local impacts, enabling more remote initiatives to be considered in offsetting local harms; and the *Voluntary Land Acquisition Management Policy* contracted the scale at which impacts were required to be mitigated. Each of these policy instruments provided the means to overcome concerns expressed at the local scale, and served to legitimise the proponent's (and wider industry) discourse around the economic imperatives of development at the broader region/state scale. The Ministerial decree that the extension applications would be subject to PAC public hearings rather than PAC public meetings ensured that determinations based on these instruments would not face merits review in the courts. This, in turn, further delegitimised the views of the community and constrained their capabilities to fully participate in decision making.

Jessup observes that it is not uncommon for environmental disputes to play out with 'localized opposition facing off against powerful proponents with access to provincial or national governments' (2013: 106). Like Jessup's case on waste facility siting, though, what is interesting in the Bulga case is the way in which the law was used strategically to prioritise and reinforce a particular scale. By scaling the conflict as overwhelmingly concerned with regional and state economic vulnerability – and concurrently framing development as a solution aligned with the 'public interest' – both the government and the proponent were able to situate themselves 'at the center of power' (Van Lieshout, Dewulf, Aarts and Termeer 2011) to facilitate the approval of the mine extension. The law was used to normalise a power imbalance which, as Monica remarked, created 'a system of ripe opportunity' rather than 'a system of integrity'.

In the Bulga case, the community's capabilities to participate in the land use decision making process were constrained on a number of levels. They lacked power due to their small population and limited resources and, because their interests were framed as insignificant, they could not adequately quantify their interests within the assessment framework. Following the introduction of the *Mining SEPP*, the environmental and social impacts of place transition, which were an important consideration in Chief Justice Preston's judgement in the original Land and Environment Court hearing, were required to be ranked below the economic benefits of proceeding with development. Even once the *Mining SEPP* amendments were overturned, place-based environmental and social impacts were still viewed as factors that could be mitigated, compensated or offset courtesy of the new *Biodiversity Offsets* and *Voluntary Land Acquisition and Management* policies. Essentially, place-based concerns were minimised within the framework for development assessment. Finally, the Minister's decision to hold PAC public hearings removed the community's power to challenge the merits of the PAC's eventual decision, as it had successfully done previously. The law was used to 'entrench and reproduce disparate power relations' (Bernal 2011: 145), which constrained the community's capabilities to effectively participate in land use decision making and to challenge the outcomes of those processes and, ultimately, their capabilities to control their local environment and future in general.

From an environmental justice perspective, the Bulga case study is not the story of a community that has been historically burdened or discriminated against. Rather, it shows how a development assessment regime can be strategically manipulated by those with power to create and prioritise certain scales, and disregard local concerns. As Jessup notes, environmental injustice occurs 'when participants, subjects or features of environmental disputes are ignored, are overlooked, or their interests are downplayed' (2015: 5). There was no doubt that the Bulga community was at a distinct disadvantage in attempting to challenge the approval of the Mount Thorley–Warkworth expansion from the outset. But while the relevant development assessment regime was arguably already tilted in favour of proponents (through the time and financial constraints imposed upon citizens who wished to participate), the multiple regulatory changes implemented following the BMPA's court victories – combined with the various discursive tactics employed by the proponent and the government – served to totally disempower the community and reinforce the might of the proponent and government. Overall, the community's capabilities of effective public participation and self-determination were constricted, which constituted environmental injustice.

Notes

1 For all documents relating to the Warkworth extension application, refer to the Department of Planning and Environment, *Major Project Assessment – Mt Thorley–Warkworth Mining Complex, Warkworth Extension Project* at http://majorprojects.planning.nsw.gov.au/index.pl?action=view_job&job_id=3639
2 *Bulga Milbrodale Progress Association Inc v Minister for Planning and Infrastructure and Warkworth Mining Limited* [2013] NSWLEC 48.
3 Ibid. at [420]–[430].
4 Ibid. at [205].
5 Ibid. at [256]–[403].
6 Ibid. at [18].
7 Ibid. at [346].
8 Ibid. at [485]–[495].
9 Ibid. at [487].
10 Ibid. at [488]–[490].
11 Ibid. at [491]–[495].
12 Commercial-in-confidence letter sent from Rio Tinto to the NSW premier, 13 May 2013, copy obtained by Lock The Gate under the *Government Information (Public Access) Act 2009 NSW* ('GIPA Act') by Lock The Gate (a national organisation of local groups concerned about coal and gas development); available online at: https://d3n8a8pro7vhmx.cloudfront.net/lockthegate/pages/1623/attachments/original/1418362106/DOC091214-09122014074819.pdf?1418362106
13 Orgill (2015) provides a legal analysis of the *Mining SEPP* 'resource significance' amendments, and argues that the amendments (as delegated legislation) were not consistent with the empowering legislation, the *EP&A Act*, and hence *ultra vires*. Overall, he concludes that the amendments 'made the mine approval legal framework incoherent, complicated and potentially unworkable' and, if maintained, would be likely to give rise to judicial review litigation (2015: 501).
14 Project documents for the 'Modification Six' application are available at: www.pac.nsw.gov.au/Projects/tabid/77/ctl/viewreview/mid/462/pac/358/view/readonly/myctl/rev/Default.aspx
15 The DGRs set out the environmental assessment requirements for a project that the applicant is required to address in their subsequent environmental impact assessment.
16 *Warkworth Mining Limited v Bulga Milbrodale Progress Association Inc* [2014] NSWCA 105.
17 Ibid. at [106]–[119].
18 Ibid. at [301].

19 However, the PAC recommended that the proponent extend such rights to those who were initially granted such rights under the repealed 2010 approval (PAC 2015a: 39–40).

20 In particular, the acquisition rights that were granted under the previous Warkworth Extension approval were reinstated (PAC 2015e: 3).

4 The Namoi catchment case study: Part 1

Coal mining on the Liverpool Plains

Introduction

Coal and coal seam gas (CSG) have both been produced in Australia for some time (ABS 2001) without attracting significant controversy. In recent years, however, the extraction of coal and CSG has increasingly encroached upon important agricultural land and natural ecosystems. Extractive industries and the government advocate that coexistence of mining, agriculture and the environment is possible, but many landholders and members of the public disagree. Nowhere has the controversy been more evident than in the Namoi catchment area in north-west NSW. The region contains some of the most productive agricultural land in the state – responsible for producing a substantial amount of Australia's food and fibre – as well as significant wilderness areas. Beneath the surface of the catchment also lies abundant coal and CSG reserves, and the region is poised to become home to some of the state's largest coal and gas projects. The case studies discussed in this chapter and in Chapter 5 concern conflict throughout the Namoi catchment region, which has increased as communities grapple with multiple extractive developments. Several developments are greenfield sites still in the early stages of exploration and assessment.

This chapter begins with an overview of the Namoi catchment. The following discussion (Part 1, Namoi catchment case study) explores the decision making processes relevant to the exploration, assessment and approval of new coal ventures in the Liverpool Plains: the Caroona and Watermark Coal Projects. Chapter 5, which is Part 2 of the Namoi catchment case study, discusses exploration activities associated with the Narrabri Gas Project towards the north of the Namoi catchment. In addition to examining the specific exploration and development decisions within the catchment, the focus in parts 1 and 2 is on the implications of state-wide regulatory reforms implemented to manage conflict over extractive development. The case study as a whole reinforces the point made in previous chapters of this book that injustice may be experienced not only in terms of the outcomes of decision making processes but also in the creation and implementation of decision making processes themselves.

As with the Bulga case in Chapter 3, the Namoi catchment case study draws on media reports, public submissions, legislation (Table 4.1), case law and development application documents, along with interview data (see Table 4.2) to

Table 4.1 Legislation, regulations and policies relevant to Part 1 of the Namoi catchment case study

Law	Function
Mining Act 1992 (NSW)	Governs the exploration and production of coal within NSW, and vests ownership of subsurface minerals in the state.
Environmental Planning and Assessment Act 1979 (NSW) ('EP&A Act')	Regulates the assessment and approval processes for major developments, such as coal mines, within NSW.
Threatened Species Conservation Act 1995 (NSW)	Aims to conserve threatened species, populations and ecological communities of animals and plants.
State Environmental Planning Policy (Mining, Petroleum Production and Extractive Industries) 2007 (NSW) ('Mining SEPP')	Provides specific criteria for the assessment of mining proposals, including compatibility with nearby land uses, any impact of transport, the efficiency of resource recovery and mine rehabilitation.
Mining and Petroleum Legislation Amendment (Land Access) Act 2010 (NSW)	Amends the *Mining Act 1992* and the *Petroleum (Onshore) Act 1991* in relation to rights of access to land for exploration.
Water Management Act 2000 (NSW)	Provides for the protection, conservation and ecologically sustainable development of the water resources of the state.
Environment Protection and Biodiversity Conservation Act 1999 (Cth) ('EPBC Act')	Governs the protection of the environment and the conservation of biodiversity at the federal level.

Table 4.2 Interviewees for the Namoi catchment case study

Name*	Role
Kylie	Agricultural industry representative
Alison	Agricultural industry representative
Robert	Agricultural industry representative
Belinda	Community member
Adam	Environment organisation representative
Susan	Farmer/community organisation representative
John	Farmer/community organisation representative
Natalie	Farmer/community organisation representative
Eric	Farmer/community organisation representative
Julia	Farmer/community organisation representative
Michael	Farmer
Daniel	Farmer
Samuel	Farmer
Charlie	Farmer/local government representative
Cathleen	Lawyer, environmental advocacy organisation
Greg	State government representative
Ian	State government representative

* All names listed here are pseudonyms.

consider perspectives on justice and fairness in the different phases of development assessment and approval. Though some interviewees spoke particularly to
developments in their local area, many interviewees also had roles which positioned them to speak about developments across the catchment; accordingly,
Table 4.2 applies to Part 1 and Part 2 of the case study.

Several extractive developments in the Namoi catchment propose to impact
upon Indigenous land and cultural heritage. While I was able to discuss these
issues informally with several Indigenous elders and community members, I was
unable to obtain consent to record interviews for the purposes of quoting individuals and so I have not included quotes from these stakeholders. However, the
perspectives of Indigenous community members on extractive development in
the Namoi catchment have been reported in the media and I was encouraged by
Indigenous stakeholders to consult these accounts to highlight relevant perspectives throughout the case study.

In summary, the Namoi catchment case study provides a wide-angled view
of a region dealing with many extractive developments, and the potential
cumulative impacts upon agriculture and biodiversity. Landholders and communities across the catchment have attempted to engage with decision making processes only to find a complex regulatory system that has not sufficiently
dealt with their concerns. During the initial exploration phase, landholders
and communities were provided with minimal opportunities for formal public
input and review, and their attempts to challenge the status quo were neutral-
ised by strategic legislative changes which disempowered those who sought
to object. Attempts to reform the law to better balance competing perspectives have done little to assuage landholder and community concerns; for the
most part, the influence of industry and the power of government has ensured
that opposition interests can be effectively sidestepped, contrary to political
assurances which had promised equitable outcomes and fair decision making
processes.

Overview: the Namoi catchment

Spanning an area of approximately 42,000 square kilometres (5 per cent of
NSW), with approximately 29,400 square kilometres under agriculture (about 5
per cent of NSW's agricultural land), the Namoi catchment is a sub-catchment of
the Murray–Darling Basin. It is located in the New England/North West region
of NSW and is home to just over 100,000 people (ABS 2016b). The catchment
includes the regional centre of Tamworth. Major towns include Gunnedah, Narrabri, Boggabri, Quirindi and Werris Creek. The region has a relatively mild climate, generally favourable volcanic soils and reliable water sources. Agriculture
is a major industry in the catchment; in 2014/15, the agricultural industry was
worth over A$960 million (gross value) (ABS 2016c), producing almost 10 per
cent of the value of broadacre crops in NSW (see Table 4.3).

The catchment also contains significant biodiversity areas, including the
3,000 square kilometre Pilliga forest (the largest continuous remnant of semi-
arid woodland in NSW) and aquatic habitats of ecological importance, as well

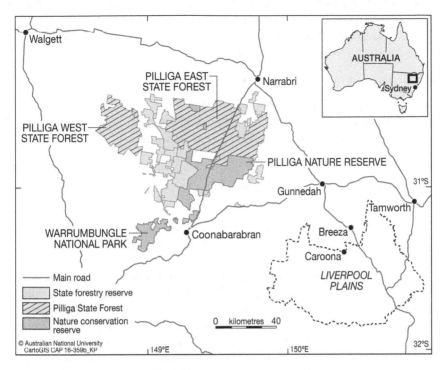

Figure 4.1 Map of the case study areas within the Namoi catchment

Table 4.3 Namoi catchment, agricultural production, gross value (2014/15)

Commodity description	A$m gross value	% NSW gross value	% Australian gross value
Cotton	149	25.9	15.6
Wheat	111	5.6	1.6
Sorghum	94	52.8	14.2
Barley	41	7.3	1.7
Pulses and oilseeds	35	11.3	2.6
Cattle and sheep (livestock products, slaughtering and other disposals)	338	7.4	1.5
Poultry (including eggs)	133.1	11.9	4.0

Source: ABS 2016b.

as endangered ecological communities and threatened flora and fauna species (Green, Petrovic, Moss and Burrell 2011).

The Gunnedah Coal Basin, containing an estimated 12 per cent of NSW's coal resources (Department of Planning and Infrastructure 2012), underlies the Namoi catchment, and coal mines have been operating in the catchment for some time. The wider New England/North West region (in which the Namoi catchment is situated) produced over 5 million tonnes (Mt) of export quality coal in 2010–11, worth A$585 million (Department of Planning and Infrastructure 2012). Existing coal mines within the Namoi catchment include:

- To the north
  - o Narrabri coal mine (near Narrabri)
  - o Boggabri, Tarrawonga and Maules Creek coal mines, each located within or adjacent to the Leard State Forest near Boggabri
- To the south
  - o Rocglen mine (near Gunnedah)
  - o Werris Creek mine (near Tamworth)

Under proposal in the southern Liverpool Plains area are the Caroona Coal Project and the Watermark Coal Project, both to be located on farmland near Gunnedah.

CSG is at a much earlier stage of development in the Namoi catchment but, in the context of increasing global energy demand, there has been a strong interest in developing CSG resources within the catchment. The NSW Department of Planning and Infrastructure states that nearly 35 per cent of the New England/North West region has been identified as having gas recovery potential (2012: 17–18), and petroleum exploration licences cover almost the entire region, including farmland on the Liverpool Plains (NSW Government, Department of Industry n.d.). Several greenfield CSG ventures are now proposed in areas within close proximity to agricultural land and sites of significant biodiversity. A major CSG development, the Narrabri Gas Project, is planned in the north of the Namoi catchment, and exploration activities have been underway for several years.

Those proposing the development of coal and gas in the catchment claim that mining is able to coexist with agriculture and, in addition, offers significant economic benefits, including alternative employment and revenue sources, as well as satisfying the growing domestic and international appetite for energy (NSW Minerals Council 2014). But communities, and agricultural landholders in particular, are concerned that the risks associated with expanding coal and CSG operations will lead to social and human health impacts, as well as a loss of arable land and the food production capacity of the region, fragmentation of rural landscapes, loss of habitat areas and cultural heritage sites, and noise and dust disturbances to adjoining properties (CCAG n.d.; Friends of the Pilliga n.d.).

Communities are especially concerned about the threats to the quality and quantity of surface and groundwater, and remain unconvinced that the risks associated with resource extraction can be adequately managed. Since European settlement, the use of irrigation for agricultural purposes has placed considerable strain on surface and groundwater resources throughout the catchment (Green et al. 2011). A series of state and federal water reforms implemented over the past

25 years to deal with water scarcity and environmental degradation has reduced the water entitlements of landholders; in response to the reforms, many landholders have implemented significant water use efficiency measures (Sharp and Curtis 2012; Green et al. 2011). The fear of landholders is that an increase in the number of extractive industries in the region will add to the demands for already-scarce water resources, placing increased burdens on aquifers, and potentially causing depressurisation that can lead to significant changes to water flow conditions (EcoLogical Australia 2011).

Complicating these concerns is the system of mineral and petroleum rights ownership in Australia, which, distinct from the position in the United States (to be discussed in Chapter 6), vests ownership of subsurface minerals and petroleum in the state. This 'dominial' system of public mineral ownership means that private landholders only control the surface of their property; the states license third parties to conduct subsurface exploration and extraction, leaving landholders with limited options to prevent access to their land once a licence has been granted.[1]

Coal on the Liverpool Plains

The Liverpool Plains Shire is located towards the south of the Namoi catchment; it is not only a 'stunningly beautiful place' (Cleary 2015), but an area well suited to agricultural production. In his reflections on life in the new colony, Peter Cunningham, former naval surgeon and early settler in the Upper Hunter Valley, described the Liverpool Plains as:

> [A]ll fine rich grassy soil without a tree, except where a small woody hill occasionally rises from the bosom of the plain to vary and beautify the prospect. In looking down upon this extensive tract from the summit of one of the overhanging ridges, the country appears to be spread out like a green ocean, of unbounded extent, with clusters of woody islands bespangling its surface.
>
> (Cunningham 1827: 159–160)

Cunningham's observations still resonate today. Picturesque mountain ranges overlook the vast plains, which contain the bulk of Australia's rare black earth and chernozem soils (Department of Planning and Infrastructure 2012). The area is suitable for year-round cropping because of the availability of surface and groundwater, reliable rainfall and a moderate temperature. The production yield of the Liverpool Plains is the highest in NSW (Department of Planning and Infrastructure 2012), reportedly 40 per cent higher than the national average (Cleary 2015). It is on these rich fertile plains where the battle over land use between extractive industries and agriculture first began in the mid-2000s.

In 2006, the NSW state government issued Coal Mines Australia Limited (CMAL) (a shelf company of BHP Billiton) with a five-year coal exploration licence (EL6505) covering 344 square kilometres at Caroona within the

Liverpool Plains.[2] BHP Billiton paid A$100 million for the exploration licence. Two years later, the state government issued another large exploration licence (EL7223) to the Chinese-owned company, Shenhua, for A$300 million for the right to explore an area of 195 square kilometres (19,500 hectares) located three kilometres from the village of Breeza and approximately 25 kilometres from Caroona. The proximity of both mines to agricultural land and the groundwater system relied upon for agricultural production led to swift opposition from the local farming community.

Initial exploration and land access

BHP Billiton maintained from the outset that it would not mine under the black soil plains, but agricultural landholders throughout the area were immediately concerned, fearing that exploration and development of coal reserves around Caroona would damage the underground aquifers that they relied upon for production. The Caroona Coal Action Group (CCAG) was established shortly after BHP Billiton obtained its exploration licence. The group states that it is not opposed to mining, but it is concerned to ensure that coal exploration and production affects neither agriculture on the Liverpool Plains nor the wider environment and community culture (CCAG n.d.). The CCAG has been at the forefront of opposition to mining on the Liverpool Plains ever since. At the heart of their concerns has been the assessment and approval processes for coal exploration and mine development.

The processes for granting exploration licences and approving exploration activities in NSW are distinct from those applicable to approving full-scale production developments (as discussed in the Bulga case in Chapter 3). In terms of exploration licences, there are limited opportunities for landholders to participate in the decision making processes concerned with their approval.[3] Further, there is minimal information available to the public on the scope and nature of potential operations because an environmental assessment is not required for the approval of an exploration licence. Proposed exploration activities, however, do require environmental assessment under the *EP&A Act* (see Box 4.1).

BOX 4.1   Assessment and approval of mineral exploration activities

For the most part, mineral exploration activities do not constitute State Significant Development (SSD) and are permissible without development consent under the *EP&A Act* (*Mining SEPP*, sch 1, cl 5). For most exploration activities, a Departmental environmental assessment takes place, which typically involves consideration of a Review of Environmental Factors (REF) – a preliminary environmental impact assessment of the proposed exploration. However, there is no legal obligation to provide the public with an opportunity to review a proponent's REF (as they would an environmental impact statement (EIS) under a full development application) under the *EP&A Act*. REFs are usually published after an exploration activity has been approved. Further, the assessment of a REF is limited to proposed exploration activities and do not contemplate impact of full-scale production.

Following the approval of BHP Billiton's exploration activities, several landholders in the Caroona exploration licence area refused to negotiate access agreements for their land, leading BHP Billiton to seek arbitration (see Box 4.2). The arbitrated access agreements determined that exploration should proceed; however some farmers continued to deny access to their land for exploration activities, and the CCAG established a blockade outside one property in July 2008 that would ultimately remain in place for some 615 days (Munro 2012; Thomson 2010).[4]

BOX 4.2   Land access agreements and arbitration

After exploration activities are approved, proponents must enter into a land access agreement with a landholder before they can go onto the land (*Mining Act* s140).

Access agreements set out the conditions of access, along with any compensation to be paid for loss or interference with the land. In the first instance, landholders and proponents directly negotiate the terms of access, but if an agreement cannot be reached within 28 days, the proponent is able to seek arbitration.

Once an access agreement is finalised, landholders cannot obstruct a proponent from accessing their land to conduct exploration activities (*Mining Act* ss139, 140, 143–153).

The landholders interviewed for this research (see Table 4.2) felt that the arbitration process was unfairly tilted in favour of proponents; Charlie described it as 'the most extraordinarily one-sided process that has ever existed'. Senior members of the NSW legal profession have also criticised arbitration under the *Mining Act*, describing it as 'outrageously skewed against landowners', 'unjust' (McCarthy 2014) and a 'pseudo-judicial farce' (Nader 2014). Of particular concern was s146 (2) of the *Mining Act*, which until recently stipulated that parties to a land access arbitration may only be represented by an agent who is *not* an Australian legal practitioner (defined as someone who is admitted to practice and holds a current practising certificate and, therefore, can give legal advice) – unless the parties consent otherwise. However, in-house corporate lawyers employed by mining companies were able to appear in arbitration matters, because the relevant professional regulations permitted in-house lawyers to perform certain legal services in the course of their employment without a practising certificate. This effectively created a loophole where legally trained persons could act for mining companies in arbitration matters, but not for landholders. Some landholders attempted to seek assistance from retired solicitors and barristers (who no longer held a practising certificate and were thus not considered 'legal practitioners'), but this was deemed problematic as it would constitute a breach of professional regulations if the former practitioner was seen to be providing legal advice (Nader 2013).

A subsequent review of the land access arbitration framework (Walker 2014) recommended that the *Mining Act* be amended to provide parties with the right to legal representation, and this reform has since been implemented (see Box 4.3).

BOX 4.3   Reform to land access arbitration processes

In 2015, the *Mining and Petroleum Legislation Amendment (Land Access Arbitration) Act 2015* (NSW) amended s146 of the *Mining Act* to provide that landholders may be represented by an agent or by an Australian legal practitioner. Other key changes of the amending Act include that exploration title holders must pay the landholder's reasonable legal costs, and gave more detailed provisions regarding the appointment of Arbitration Panel members.

Recent changes to the legal profession laws also now require all in-house lawyers to retain a practising certificate, which – if the *Mining Act* had not been changed – would have precluded them from acting for mining company employers in land access arbitration matters in any event: see *Legal Profession Uniform Law 2014* (NSW).

During the period of the case study described in this chapter, however, these amendments had not yet commenced.

The processes for the assessment and approval of exploration licences and activities, and the arbitration process concerning access agreements, created disadvantages for landholders and members of the community. In other words, there was a distinct lack of procedural fairness (see Chapter 2), because the public were not provided with sufficient notice or adequate information concerning the granting of exploration licences and activities. In addition, the public were not given opportunities to participate in, let alone review or challenge, decisions regarding exploration licences and activities. Landholders, in particular, expected to have a much greater involvement in the process of determining access for exploration. While the arbitration process provided some opportunity for them to negotiate the terms on which proponents could access their land, the disparity in representation during arbitration processes and the narrow framework within which objections could be raised limited what they could do. In

Caroona, the only option for landholders was to challenge the validity of their arbitrated access agreements in the Mining Warden's Court, which some land-holders ultimately did.

Backed by funding from the Australian Farmers' Fighting Fund, one group of farmers argued that their respective access agreements were invalid because the proponent had failed to adequately notify the landholders' banks (as mortgagees of the farms) of their intention to prospect for coal.[5] While the Court agreed that mortgagees ought to have been notified of the intention to conduct explo-ration activities, the Court held that the failure to serve notice did not invali-date the access agreements. Mining Warden Bailey observed that the *Mining Act* 'has obviously deliberately taken away from landholders the right to object to an exploration licence'.[6] He further noted that, under the legislation, arbitrators (as well as the Court) did not have the power to override exploration rights granted by the minister.[7]

An attempt to shift the balance of power: the Supreme Court appeal

In early 2010 two landholders, the Browns and the Alcorns, appealed their cases to the Supreme Court of NSW, arguing that the Mining Warden's Court had erred in not deeming the agreements invalid because of BHP Billiton's failure to notify mortgagees of the potential impacts upon the mortgaged farm properties:[8]

- The Browns were seeking organic certification for their property, and were concerned that access by BHP Billiton would jeopardise their certification through the potential contamination of aquifers and soil.
- The Alcorns, who operate a Limousin cattle breeding stud, were concerned about contamination of their lagoon and the potential for their cattle to be poisoned.

Justice Schmidt of the Supreme Court agreed, finding that there can be many landholders who have a legitimate interest in a property (including mortgagees, native title holders and lessees), and that they must each have an opportunity to be notified of proposed exploration as well as a right of appearance at any arbi-tration hearing concerning the land. In light of this, Justice Schmidt found that Mining Warden Bailey 'did not consider that the concerns lying at the heart of the dispute between the parties, namely conditions minimising the risk of con-tamination and damage to the land which provides the security for the funds advanced by the mortgagees to the plaintiffs, might also be of concern to the mortgagees'.[9] Both the access agreement determinations and the decision of the Mining Warden's Court were overturned. Power shifted to the landholders as the law legitimised their claims.

However, the landholder's victory was short-lived. In June 2010, only weeks after the Supreme Court judgement, the state government introduced what the NSW Minerals Council deemed to be a 'sensible remedy' (Robins and Cubby 2010) for the decision: the *Mining and Petroleum Legislation Amendment (Land*

Access) Act 2010 (NSW). The legislation introduced specific amendments to the *Mining Act* to classify those interest holders not in immediate possession of the land (e.g. mortgagees, such as financial institutions) as 'secondary landholders', with whom there is no obligation to enter into an access arrangement before commencing exploration. Critics argued that the amendment was being used to give priority to mining interests (Robins and Cubby 2010); the NSW Minerals Council (as representatives of the wider industry) had been particularly scathing, in the media, of the Supreme Court's decision and strongly lobbied for reform. Then-CEO Dr Nikki Williams claimed that the Supreme Court decision would result in:

> [A] complicated, convoluted bureaucratic nightmare that will make it impossible to do business in NSW . . . the NSW Government, the Opposition and businesses should be very worried. They need to take action now to avert this job destroying, wealth destroying minefield.
>
> (Condon 2010)

Landholders believed that the mining industry had played a central role in influencing the government to quickly change the law after their Supreme Court victory. As one landholder interviewed for this research, Charlie, stated:

> We got millions of dollars of support from the farming community to take the miners to court and we won. And what happened when we won? Six weeks later the New South Wales government retrospectively changed the law and basically flushed that money down the toilet. [W]hen you look at it, that's injustice . . . [the government] can just say [to industry] 'don't worry about it we'll just change the law'.

The formal legal act of changing the *Mining Act* certainly supported wider industry framing of exploration as essential to job security and economic development. Much like the *Mining SEPP* amendments discussed in the Bulga case in Chapter 3, the amendments implemented here smoothed the path to development approval for proponents. The industry advanced scale frames around the economic imperatives of development at state and regional levels in lobbying for reform, and these were echoed in the subsequent frames deployed by the government to justify their reforms. In particular, the then-Minister for Mineral and Forest Resources, Ian Macdonald, stated that the amendments to the *Mining Act* were necessary in order to 'facilitate exploration for the benefit of the people of NSW', and to remove 'impossibly bureaucratic, lengthy and costly delays' for all parties (Robins and Cubby 2010). This effectively disregarded the interests of landholders and bolstered industry claims about the need to pursue exploration for the public benefit.

Meanwhile, tensions had also started to grow in nearby Breeza, with Chinese-owned Shenhua's Watermark Coal Project emerging as another potential key development in the region. However, Shenhua took a different approach to BHP

Billiton to gain access to land. Rather than challenging landholders for access rights under the *Mining Act*, the company began purchasing properties at well above market rates. By mid-2010, Shenhua had acquired more than 25 properties for an estimated total of A\$130 million (Wahlquist 2010). By mid-2011, this had increased to 43 properties, comprising at least 14,700 hectares (approximately 75 per cent of the exploration licence area) (Munro 2012), with a total spend of A\$213 million (Bita 2011). Some properties sold for more than ten times their previous sale value.

There was widespread public criticism of the federal government's decision to allow such rapid foreign acquisition of Australian farmland, but Shenhua's reported A\$213 million expenditure fell short of the A\$231 million threshold which triggers investigation by the Foreign Investment Review Board (Madden 2011). Some members of the public blamed the state government for granting the exploration lease in the first place (Myers 2010), and landholders interviewed spoke of division in the community as well as acknowledging that some landholders were left with no choice but to sell. Many of the properties acquired were leased back to farmers but, in the meantime, the purchases provided Shenhua with the opportunity to grant itself access rights for exploration activities and avoid the issues encountered by BHP Billiton in Caroona. Essentially, Shenhua's overwhelming economic power ensured that opposition encountered at the exploration stage could be easily organised out.

The conflict generated by the Watermark and Caroona coal exploration projects saw development planning and assessment emerge as one of the key issues at the 2011 state government election (Jopson 2011). Following a landslide victory, the incoming O'Farrell Liberal/National Coalition government declared a moratorium on the granting of new exploration licences for coal and gas projects and announced that additional conditions would be imposed on both Shenhua and BHP Billiton's exploration licences which would prohibit both open cut and longwall mining on or under the areas of the alluvial floodplain and the black soil plains (Hartcher 2011). The new government also promised to overhaul the planning legislation at the heart of land use conflicts. Central to their reform package was the removal of the controversial Part 3A (*EP&A Act*) system which granted wide ministerial discretion to approve development projects (as discussed in Chapter 3, Box 3.2) and the development of a *Strategic Regional Land Use Policy* ('SRLUP') to 'strike the right balance' between agriculture and mining (Hazzard 2011).

The Strategic Regional Land Use Policy: finding a balance or playing politics?

The government introduced the SRLUP reform package in September 2012; it included a range of new initiatives, including codes of practice for CSG drilling; the establishment of a Land and Water Commissioner to provide independent advice on mineral and gas exploration; an Aquifer Interference Policy to manage water use and conflict; standardised land access agreements; and a 'Gateway' process where an independent Mining and Petroleum Gateway Panel would assess the impact of extractive developments on 'Strategic Agricultural Land' (SAL) (see Box 4.4) (Hazzard 2011).

BOX 4.4   Strategic Agricultural Land

Strategic Agricultural Land (SAL) is land considered to be highly productive and possessing unique natural resource characteristics. SAL is divided into two categories:

1.   Critical Industry Clusters (CICs), which exist where there is a high concentration of productive agricultural industries that contribute to the identity of the region and provide significant employment opportunities, such as the viticulture and equine industries in the Hunter Valley. CSG activity is prohibited in mapped CIC areas; while coal mining is permissible subject to assessment by the Gateway Panel.
2.   If high-value agricultural land is not included in a CIC, it may still be mapped as Biophysical Strategic Agricultural Land (BSAL). BSAL is identified as productive agricultural land, containing high-quality soil and water resources. Coal and CSG developments proposed on BSAL must undergo Gateway Panel assessment.

SAL was identified through a process of mapping contained in Strategic Regional Land Use Plans, with input from written public submissions and public forums. The first SRLUPs released in 2013 covered the Upper Hunter and the New England North West regions and included the Shenhua and BHP Billiton exploration sites. Further mapping for the remainder of the state was completed in 2014 (Department of Planning and Infrastructure 2014).

Strategic spatial planning has been increasingly adopted as a method to govern difficult policy spaces, providing a means to coordinate regulatory responses across multiple departments and account for the cumulative impacts of development. Such spatial planning 'identifies the capacity of land for particular forms of development, and sets long-term strategic goals and a blueprint for development of a particular region' (Owens 2012: 117). With one of the central planks in the O'Farrell government's election campaign a promise to protect prime farmland and balance competing land uses, agricultural communities were hopeful that the SRLUP would provide a new direction in strategic planning within NSW. Throughout the public consultation process for the SRLUP, many farming and community groups reminded the government of its pledge to create exclusion zones for high-value agricultural land (Nicholls 2011; Cubby and Nicholls 2012).

'Fighting with one hand tied behind our back'

The initial consultation draft of the SRLUP released to the public had provided the Gateway Panel with the ability to refuse development where it would be likely to cause significant detrimental impacts upon SAL. However, during the consultation process for the SRLUP, it soon became apparent that industry groups, such as the NSW Minerals Council and the Australian Petroleum Production and Exploration Association (APPEA), were actively lobbying the government against exclusion zones for agriculture (Cubby and Nicholls 2012). One of the landholders interviewed for this research, Julia, observed:

> APPEA sometimes Tweet things like 'We have 54 lobbyists in Canberra today'. And there's us that have been waiting and have to go through the correct channels, through our local member – so access to Ministers and government is pretty remote for us . . . when you think of the open access that the resource industry has to parliament, it's just outrageous.

Charlie believed that consultation over the SRLUP was actually 'about playing politics'; Samuel noted that 'it's not about governance, it's about slogans during elections, and it's about doing deals . . . [we're] fighting with one hand tied behind our back'.

Much to the ire of farming groups and agricultural communities, the SRLUP that was finally implemented limited the power of the Gateway Panel to being able to issue an unconditional certificate, or a conditional certificate which recommends project modifications or further studies before an application for development assessment under the *EP&A Act* can proceed. Thus, 'no matter how questionable the application' (Sherval and Graham 2013: 178), the Gateway Panel is not able to refuse an application outright. This caused farming and community groups to protest that the Gateway process was a 'toothless tiger' that would offer no protection to high-value agricultural land (Foley 2013).

As a result of the apparently successful efforts of the industry lobby to have the SRLUP amended in its favour, landholders – as well as community members more generally – felt that local desires for the landscape and place-based values were not given the same priority as the profit potential of resource extraction within the final SRLUP framework. Kylie captured the view of a number of interviewees for this case study:

> The Government committed when they came in that they would change the old Part 3A process and that they would overhaul the planning system, and we would have this fantastic new regulatory framework that would provide much more transparency and accountability and also provide communities with some assurances. So we went through all that consultation process for [the SRLUP] . . . yet no-one can actually demonstrate where the Government has used the feedback from communities.

Even one of the state government representatives interviewed complained that the final SRLUP had done little to change what was perceived to be an unequal system:

> O'Farrell came in on an election promise . . . there was something rotten with the planning system and he was going to make it more open and what have they done? They've done the exact opposite!
>
> (Ian)

Academic commentators, in particular, argued that the government adopted a narrow and rigid approach to identifying SAL (see Box 4.4) within the SRLUP (Owens 2012; Sherval and Hardiman 2014), noting that the suite of reforms represent a 'dangerous shortcut' in planning regulation, an attempt to 'depoliticize the issues at stake, and provide a reductionist and once-off solution to a complex political question' (Owens 2012: 113–114). Sherval and Graham argue that the SRLUP 'further entrenches imbalance by rendering it impossible to undertake genuine evaluation' (2013: 178) because, in spite of the suggestion that 'an open and transparent consultative approach has occurred', it still remains 'relatively easy' for companies to obtain approval for extractive activities (Sherval and Hardiman 2014: 199). Several interviewees expressed disappointment regarding the consultation process for the SRLUP and the outcomes of the process:

> I think on the surface the Government could claim some outcomes [of the SRLUP] included better transparency and a consistent process and supposedly independent decision making . . . [but] I actually think that's window dressing . . . the outcome they wanted was really about a perception of those things being achieved rather than really genuinely engaging the community.
>
> (Kylie)

> If you don't have any control over the outcomes, then it doesn't matter how much advice you give, it can be ignored . . . [it is] a disempowering thing for the community to be part of that.
>
> (Eric)

As argued in Chapters 1 and 2 of this book, a failure to adequately account for place-based perspectives in land use decision making practice can marginalise communities and constrain their capabilities to participate in decision making as well as constraining their capabilities to control their local environment and future more generally. Groves's (2015) work, in particular, highlights that place attachment can become 'colonised' and controlled by those in power, who establish the parameters within which place-based concerns can be accommodated. Where those boundaries are insufficient to account meaningfully for place-based concerns, the very identity of citizens can be overlooked or ignored. As Sherval and Graham observe, the problem 'is not that land may be profitable; the problem is the land is *not only* profitable' (2013: 178, emphasis added). By devaluing local perspectives on place and permitting development in all scenarios, the SRLUP created a disjuncture between community identities and the broader economic logic of the assessment regime, arguably morphing the process of landscape capacity evaluation with political decision making (Owens 2012). Simply put, the rhetoric of reform was different to its realities, and landholders felt disrespected and disadvantaged as a result. As Gross highlights, people attach 'great importance' to their treatment from government when a new process or policy is introduced, and perceptions of poor treatment 'can have a serious impact on their personal well-being' (2014: 108). Even where consultative processes exist, a failure to meaningfully recognise particular perspectives expressed within those processes can be perceived as an injustice.

The independent Gateway Panel was established in late 2013, by which time Shenhua had already lodged an application to develop the Watermark Coal Project (lodged in late 2012). BHP Billiton sought to commence its application process in early 2014 (see Box 4.5). The government decided that both developments would undergo Gateway Panel assessment before progressing to full merit assessment under the *EP&A Act* (see Chapter 3, Boxes 3.2 and 3.3 for information on the merit assessment process), providing an opportunity to test the rigour of the Gateway process.

BOX 4.5   Applications to develop coal mines on the Liverpool Plains

In late 2012, Shenhua submitted a full application for SSD consent under Part 4 of the *EP&A Act* to develop its 30-year Watermark open cut coal mine. The project application promised vast economic benefits for the region and the state, including A$1.5 billion worth of royalties over the mine's lifespan and the creation of hundreds of jobs during the construction and operation phases of the mine (Shenhua n.d.). Approximately 159 million tonnes of coal would be produced from the mine, the majority being semi-soft coking coal with the remaining 15 per cent thermal coal for power generation (Department of Planning and Environment 2014a).

In early 2014, BHP Billiton commenced the application process for its Caroona underground coal mine, submitting a request for Secretary's Environmental Assessment Requirements (SEARs) (previously called Director-General's Requirements, or DGRs) so that it could prepare a full application and EIS. The preliminary environmental assessment submitted for the purposes of obtaining the SEARs indicated that the project would involve underground longwall mining over 30 years to extract approximately 260 million tonnes of coal, employing up to 400 people at its peak operational phase (BHP Billiton 2014).

'There is no gate in the Gateway'

The Mining and Petroleum Gateway Panel is comprised of independent scientific experts from agricultural science, hydrogeology and mining and petroleum development. Members are appointed for a three-year term. The Gateway Panel assessment is 'an upfront, strictly scientific assessment' of the agricultural impacts of mining and CSG developments on SAL. There are no opportunities for public consultation in the Gateway assessment process.[10]

Despite having progressed past the point in the process where a Gateway Panel assessment would ordinarily be triggered,[11] in late 2013, the Minister for Planning and Infrastructure sought advice from the Gateway Panel on the potential impacts of Shenhua's Watermark development on BSAL identified within the project area. Following BHP Billiton's request for Environmental Assessment Requirements in early 2014, the Caroona project was also required to undergo Gateway Panel assessment before the development of a full application and EIS (see Box 4.4).

The Gateway Panel handed down its report in January 2014 on Shenhua's proposed Watermark project and expressed several concerns, noting:

- The EIS submitted for the project and Shenhua's agricultural impact assessment had only been conducted for the disturbance area rather than the entire project area (Mining and Petroleum Gateway Panel 2014a). This meant that impacts on surrounding BSAL soils were not considered. The panel insisted that the agricultural enterprises and systems of the adjoining locality should have been studied.
- The Gateway Panel found that the presence of saline water in backfilled areas and the final void pit lake posed a 'significant long-term risk to water quality in surrounding creeks and alluvial aquifers', and that it was 'reasonably foreseeable' that the mine would make some water unsuitable for agricultural use (Mining and Petroleum Gateway Panel 2014a: 3).

Overall, the panel found that Shenhua's methodology had underestimated the impact upon agricultural land. The panel advised that the project should not proceed until a more detailed assessment of BSAL was undertaken. The panel recommended further water modelling and consideration of changes to the mine plan. However, because the minister had only asked the Gateway Panel to provide advice on the Watermark project, the views of the panel were deemed to be merely another factor that the decision maker could take into account. The Planning Assessment Commission (PAC), as decision maker, stated that the Gateway Panel's advice did not 'technically apply' to their assessment of the project (Planning Assessment Commission 2014a: 10; the PAC's assessment of the Shenhua Watermark project is discussed in further detail in the following section).

For BHP Billiton's proposed Caroona coal mine, the Gateway Panel again expressed significant concerns about the methodology used to identify BSAL and highlighted deficiencies in the water modelling (Mining and Petroleum Gateway Panel 2014b). Even on the limited data, the panel considered that the project would have 'direct and significant' impacts upon agricultural land and ground-water, particularly subsidence across irrigated paddocks, and groundwater levels. Despite the panel's adverse findings, it only had the ability to issue a Conditional Gateway Certificate, meaning that the proposal could still proceed to develop-ment assessment (*Mining SEPP*, Reg 17H).

The outcomes of the Caroona and Watermark Gateway assessments confirmed landholder and agricultural industry criticisms that the Gateway process was 'powerless' and a mere 'legislative hoop' (Hannam 2014c). The then-President of NSW Farmers, Fiona Simson, remarked that although the SRLUP had enabled the mapping of valuable BSAL and water resources, the Gateway process had let down farmers: 'there was nothing provided to protect them . . . [i]t makes no sense to map our best natural assets yet still grant exploration licences over them' (Simson 2014). One farmer interviewed in the media quipped that '[i]t's not a Gateway, it's an autobahn – get there as fast as you can' (Hannam 2014c). Interviewees for this case study agreed:

> The Gateway process is basically just extra steps to a final determination in favour of resource extraction.
>
> (Julia)

> The Gateway process is an exercise in teleology, in language. It is a pack of lies.
>
> (Samuel)

In the eyes of landholders, while various frames had been advanced by the gov-ernment that the SRLUP and the Gateway process would protect valuable agri-cultural land and account for local concerns about key agricultural enterprises, in reality, the processes themselves did not live up to these assurances. In particular, the ability for the PAC to reject the Gateway Panel's findings was considered by landholders as a failure to recognise and respect their interests, and a further example of the combined power of industry and government working to limit their capabilities.

Despite receiving a Conditional Gateway Certificate and preparing a prelimi-nary environmental statement in order to initiate the process of development assessment, BHP Billiton did not immediately submit a full application and EIS for its Caroona project. However, Shenhua's Watermark project was ready to pro-ceed to the next phase of development assessment under the *EP&A Act*, involv-ing assessment by the PAC (see Chapter 3, Box 3.3).

Assessment of Shenhua's Watermark development application

When the Minister requested the Gateway Panel's advice on the Shenhua project in late 2013, he directed the PAC to undertake a review of the project (including conducting a PAC public hearing – see further Chapter 3, Box 3.4) and take account of the Gateway Panel's advice in their review. This required landholders and other stakeholders to engage in making written submissions to the EIS and verbal submissions to the PAC to challenge the development.

During the public exhibition period, the Department of Planning and Environment received over 130 submissions on the project, the majority expressing concerns, which included:

- the likelihood of increased noise and dust pollution on the community and the environment;
- potential impacts upon the local koala population; and
- potential impacts upon aquifers and agricultural production on some of Australia's richest soils – though mining under the areas of the black soil plains had been prohibited under the exploration licence, there were fears that excavation of the adjacent ridges would impact groundwater quality and quantity.

The project continued to attract significant opposition throughout the PAC review. The CCAG and other objectors particularly criticised the groundwater modelling by Shenhua in its EIS, basing their criticisms on assessments they had commissioned (Earth Systems 2013) and evidence from the independent Namoi Catchment Water Study (see Box 4.6). The NSW Irrigator's Council and the CCAG jointly engaged the University of New South Wales' (UNSW) Water Research Laboratory to conduct a further peer review of the groundwater modelling in Shenhua's EIS for the purposes of the PAC public hearing. The UNSW study found that Shenhua's modelling did not test all geologic scenarios, nor did it adequately represent the complexity of the groundwater system and aquifer connectivity (Water Research Laboratory 2014). The Water Research Laboratory study concluded that the groundwater predictions in Shenhua's EIS were 'subjective, incomplete, contain bias and are not adequately supported by the available data' (Water Research Laboratory 2014: 22).

BOX 4.6   Namoi Catchment Water Study

The Namoi Catchment Water Study was previously commissioned by the state government in 2010 in response to community concern regarding the impacts of mining and gas exploration on water resources throughout the catchment. The study modelled various scenarios of coal and gas development to test the impact upon groundwater levels and flows, but its interpretation became an issue of conflict in the community. Government and industry took the report as proof that coal and gas extraction would only have a low impact on water resources at a regional scale; however,

stakeholder groups were critical that the study's findings had been misin-
terpreted and understated (Farr 2012). In particular, stakeholder groups
argued that the study had found that cumulative development would greatly
increase local scale risks to water resources, and some groups believed that
the models which outlined more intense development scenarios demonstrated unacceptable risks in some areas (Hardy 2012).

All documents relating to the Namoi Catchment Water Study are available online
at: www.resourcesandenergy.nsw.gov.au/info/namoi-catchment-water-study

The PAC also commissioned an independent review, which raised similar questions about the design and methodology employed in the Shenhua's groundwater
model. The review noted that the model's outcomes were inaccurate and did not
'properly represent the impacts on the groundwater systems that would arise from
the project' (Planning Assessment Commission 2014b).

In its review report, the PAC acknowledged that uncertainties had been
expressed by opponents about Shenhua's water modelling. Nonetheless, it
concluded that the project was approvable. It was noted that the PAC review
was 'only one step in the assessment process' (Planning Assessment Commission 2014a: 14), and recommended additional water modelling be carried out to
resolve uncertainties before the project's final determination. The PAC considered that the remaining issues raised by objectors were acceptably managed in the
proponent's mine plan. In particular, it found that the proposed location of the
mine on the slopes adjacent to the black soil plains would not have a significant
impact on agricultural productivity. As noted earlier, the PAC acknowledged the
concerns raised by the Gateway Panel as to the proponent's conservative estimate
of BSAL land within the mine site, but concluded that BSAL is only one measure
of agricultural value, and that, in any event, the Gateway process ultimately did
not apply to this project. The loss of agricultural land predicted in the EIS was
deemed 'relatively small' and not likely to have a significant impact on agricultural productivity at the wider regional scale.

Another concern for consideration by the PAC was the potential impact upon
the local koala population, which was already suffering from the effects of prolonged drought; concerns had been raised that the mine development would
result in a loss of habitat that would cause a significant population decline. The
PAC found that more work needed to be done to monitor the koala population
but found that translocation of koalas from the mine site was possible.

In its review report, the PAC made 25 recommendations concerning further research and water modelling. Shenhua and the Department of Planning
and Environment were given the opportunity to respond to the PAC's findings
(Department of Planning and Environment 2014b). Most of the PAC's recommendations were rejected by the proponent and the Department on the grounds

that they were outside the scope of the PAC's authority, not relevant to the assessment of the project or not feasible to implement.

The project was reconsidered by the PAC for final determination in late 2014. A PAC public meeting was held in which the impacts of the development upon groundwater resources, biodiversity, agriculture, noise and air quality, and the local koala population were raised. Impacts upon Indigenous cultural heritage were also brought to the PAC's attention (see Box 4.7).

BOX 4.7   Gomeroi People's heritage

The Liverpool Plains region is rich in Indigenous heritage. The proposed Shenhua Watermark project is located within Gomeroi (also called Kamilaroi) country, home to the Gomeroi people. Shenhua's EIS had identified that there would be disruption to several Gomeroi archaeological sites in the ridges to be excavated, in particular two 'grinding grooves' sites which were traditionally used for sharpening weapons, but the Department of Planning and Environment had previously advised that the mine would be unlikely to have a significant impact upon Aboriginal cultural heritage values, and that the grinding grooves sites could be relocated (Planning Assessment Commission 2015g). At the public meeting, a number of Indigenous community members spoke of the significance of the sites, and expressed the view that the grinding grooves had to be protected in perpetuity *in situ* (Planning Assessment Commission 2015g). As one traditional owner explained to the media, the grinding grooves sites:

> [A]re just like our burials, that's how significant they are to us. Them grinding grooves are the warrior's grinding grooves. They are the warriors, the ancestors who protected us. It is our job to protect them now and our future.

(Fuller 2014)

Despite all these community concerns, in January 2015 the PAC approved the Watermark mine. The PAC found that:

- Indigenous cultural sites could be relocated with little impact upon cultural heritage;
- the impacts to the local koala population could be managed by a Koala Technical Working Group;
- the proposed biodiversity offset package – which did not provide 'like-for-like' offsets – was consistent with the recently implemented *Biodiversity Offsets Policy for Major Projects* (see Chapter 3, Box 3.9);

- the mitigation and acquisition proposals to deal with noise and dust disturbance were consistent with the criteria under the recently implemented *Voluntary Land Acquisition and Management Policy* (see Chapter 3, Box 3.10);
- the economic benefits to the broader community would far outweigh any negative local social and environmental impacts. The PAC reiterated that the location of the mine on the hills and slopes adjacent to the fertile black soil plains would 'not preclude the continuation of significant agricultural production' across the region (Planning Assessment Commission 2015g); and finally;
- that the mine would have a limited impact upon groundwater quality and quantity. Shenhua's revised modelling, which the Department had reviewed by their own groundwater expert, once again maintained that the drawdown on aquifers would be limited, and that the groundwater impacts of the project were acceptable (Department of Planning and Environment 2014b). The PAC concluded that the groundwater impacts of the project had now been modelled correctly and were manageable, subject to ongoing monitoring (Planning Assessment Commission 2015g).

Following the PAC's final determination, landholders and the wider community felt that many of their concerns remained unanswered. Specifically, they believed that there was significant uncertainty around aquifer connectivity between the ridges and the plains (Foley 2015a), as well as around the management of environmental impacts, Indigenous cultural heritage and the local koala population. Landholders and community members interviewed spoke of the nature of public participation during the PAC consultation processes and, in particular, their inability to challenge the claims of the proponent following the PAC hearing and the merits of the PACs decision,[12] which restricted their capabilities to participate effectively in the land use decision making process and, ultimately, determine their future.

'You are repressed while being tolerated'

As with the Bulga case discussed in Chapter 3, interviewees for this case study emphasised the difficulties associated with obtaining the necessary technical advice to adequately assess Shenhua's EIS within the two-month window that was available for public participation:

> The Department of Planning . . . and the proponent, they spend quite a number of years or months in the lead up to getting these projects where they think they are satisfied . . . [but] for the community to try and engage with that process is just such an enormous feat . . . I mean I deal with communities all the time that just decide not to. They just say: 'this is far too hard, there is no way in 28 days or even two months I am going to be able to read the volume of that material given I've got three kids and two jobs'.
>
> (Cathleen)

It is very technical, you have got to read a very technical thing and there are parts of the community that couldn't even read that – and then you have got to write something.

(Alison)

These complaints were echoed by local Indigenous peoples. Shenhua representatives complained that the Gomeroi Traditional Custodians had been inconsistent in expressing concerns about the significance of the grinding grooves sites (ABC Online 2014), but a representative of the local Aboriginal Land Council claimed that elders had struggled to express their views within the parameters of the consultation processes:

To hand a person an Aboriginal cultural heritage plan on a disk that when a person doesn't have access to computers and think that is an agreeable form of consultation or handing them a 400 page document and asking them to read that, review it and provide a response and hang on we want that in 28 days – thank you very much – that's clearly not enough.

(Fuller 2014)

A number of interviewees also argued that the financial costs of obtaining the necessary data to challenge the proponent's claims, such as the study provided by the Water Research Laboratory, was a significant obstacle. Adam noted:

The mines pay the consultant $100,000 or whatever to deliver [the EIS] and we're expected to go out there in our own time and at our own expense and come up with counter arguments that are equivalent to what they've been paid a lot of money to do.

Susan highlighted the constraints for obtaining finances:

We get our money by lamington drives [fundraising events], but we have to shell it out in huge chunks to get the scientific stuff to really back up what we're saying, so we're at a distinct disadvantage.

In addition to the financial outlays, many interviewees highlighted the considerable personal costs borne by those who were active in challenging the developments, including time away from paid employment, and the social and emotional toll:

We've actually had people taken away from their farming businesses . . . to focus on an issue wholly and solely to protect their livelihood and the sustainability of their business long term.

(Kylie)

It comes to a huge amount, we just don't cost it . . . It is never quantified . . . [and] the mental pressure . . . that's a cost we don't measure.

(John)

There's a lot of fatigue now I think.

(Daniel)

My income is severely restricted, I spend a lot of my income actually on this business, not my business, but coal.

(Susan)

[M]y work has suffered enormously. I'm going down the gurgler financially.

(Adam)

In light of the standards identified for procedural fairness, the experiences of interviewees indicate asymmetry in information provision, which impacted their capabilities to participate effectively in the PAC's decision making process. The financial and personal costs shouldered by individuals and communities to defend their livelihoods and way of life was viewed by interviewees to be a significant injustice. Charlie summarised the views of many:

> [Mining companies] are paid to do it because that is their business whereas we're actually away from our businesses, away from our farms, away from all the things that we should be doing to keep our operations going. And the government don't take any of that into account . . . When you think about the impost on families, in people's lives, in their businesses, in their whole psyche, in their spirits, it's the most extraordinarily unjust thing.

Moreover, the failure of industry and government to acknowledge these burdens evinced a lack of recognition. As Samuel put it, landholders and the community in general were 'repressed while being tolerated' within the PAC's public participation processes, and considerably limited in their abilities to direct their future. The failure to recognise the unique circumstances of the Gomeroi people also translated into a structural inequity in their capabilities to participate effectively in the land use decision making process.

'It's a law for them and a law for us'

Another complaint of landholders and other stakeholders interviewed was the lack of opportunity to review and challenge the updated modelling prior to the PAC's final determination, leading community members to characterise the PAC as 'just a rubber stamp' (Natalie). Interviewees characterised the PAC's consultation processes as 'failing communities' and 'extremely flawed' due to 'empty consultation':

> I think [the panel] were more engaged about managing the process and the perception of it than actually the level of detail. I mean, I don't think that they really asked questions of people that were presenting that were well researched, that understood the community's issues . . . in terms of the way that they asked questions of the proponent, I don't think that they were probing enough.

(Kylie)

Dr Ian Acworth, from the Water Research Laboratory, noted in a public statement that the proponent and decision makers had failed to consult previously published peer reviewed studies:

> I published [in the *Australian Journal of Earth Sciences*] in 2009 a paper that shows . . . observations of fracturing and leakage in the clays . . . from a site on the Plains very close to the proposed mine . . . My principal concern is that this sequence of silt and clay dominated soil will leak badly when drained – as proposed adjacent to an open mine, or, for that matter, by too much abstraction by the agricultural sector. These concerns do not seem to be getting any consideration – despite being in the public domain as published papers . . . I have been working on this part of the Plains for 20 years yet I have not been approached by any one of the . . . [PAC] review groups . . . My concern is that the conceptual model is simply wrong. If this is the case then the groundwater models will all have given the wrong answer – no matter how good the various professionals are.
>
> (Water Research Laboratory 2015)

The Ministerial direction to hold a PAC public hearing (see Chapter 3, Box 3.4) precluded objectors from seeking a subsequent merits review of the project before the Land and Environment Court (*EP&A Act* s 23F). Even though a public meeting was held as a part of the PAC's eventual determination process, because a PAC public hearing had already occurred during the initial PAC review, the merits of the PAC's final decision could not be challenged. The CCAG argued that, while the PAC had provided a mechanism for the community to raise their concerns, it 'failed in providing a pathway to resolving these issues' (CCAG 2015a, 2015b). Susie Lyle, chair of the CCAG, remarked:

> Numerous times we have put forward independent evidence that the mines should not go ahead in our region and every time we do so the mining company gets another chance to change their development application, the goal posts keep moving.
>
> (Alden 2014)

The CCAG noted several matters discussed at the final PAC public meeting that were not open to further review or challenge by the community, such as the definition of 'floodplain' used by Shenhua in its EIS. Consistent with the conditions of the exploration licence, which prohibited 'longwall mining underneath the deep alluvial irrigation aquifers and the floodplain and open cut mining anywhere on the floodplain' (Planning Assessment Commission 2015g), Shenhua maintained that mining activity would be limited to ridges and slopes adjacent to the plains. However, with no definition of floodplain prescribed in the licence, the Division of Resources and Energy (within the Department of Trade and Investment, Regional Infrastructure and Services) provided its own definition, which defined a floodplain as 'an area of low-lying, nearly flat plain

adjacent to a river, formed mainly of river sediments and subject to regular flooding' (Planning Assessment Commission 2015g; Foley 2015b). But this definition was narrower than gazetted definitions of the term,[13] leading to assertions
by the community that, as defined, the proposal would not cover any areas of
'floodplain' as commonly understood and implemented by farmers under water
sharing plans (Foley 2015b).[14]

The lack of a 'real right of rebuttal' to challenge the merits of such decisions
was considered by interviewees for this case study to be a 'massive inequity':

> PAC hearings are not worth two bob. There's no legal argument about merits
> for or against. They're just an opportunity for people to vent their spleens
> about whatever they don't like.
>
> (Adam)

> The practice of referring matters for public hearing means that you don't
> have equitable access to the system of challenge and testing proponent's
> material . . . the fact that the public hearing removes the merit appeal rights
> is just a bit silly really . . . they're like apples and oranges. A merit appeal
> and a Planning Assessment Commission public hearing are just entirely
> different procedures and entirely different functions, and they shouldn't be
> mutually exclusive.
>
> (Cathleen)

Both the conduct of the PAC's public consultation processes, and the inability to review its findings, raise further issues of procedural fairness. As Charlie
highlighted:

> It's a law for them and it's a law for us . . . There is absolutely no procedural
> fairness in any of this process whatsoever. It is the most unjust one-sided
> thing and it is a disgraceful abuse of human rights.

Arguably, the decision to hold a PAC public hearing – which made it impossible to seek merits review following the PAC's final determination – also raises
issues of misrecognition. The proponent was provided with several opportunities
to respond to matters raised at the final PAC public meeting yet landholders and
the community had no such ability to challenge the merits of particular decisions (e.g. the floodplain definition). Their interests were able to be legitimately
ignored under the conditions of the regulatory framework, which fundamentally
impeded their capabilities to participate in the land use decision making process.

Attempts to challenge the Shenhua project PAC decision

One community group – Upper Mooki Landcare – did, however, seek judicial
review of the PAC determination, commencing proceedings in the NSW Land
and Environment Court to challenge the approval of Shenhua's Watermark mine

on an error of law.[15] The group argued that the PAC did not properly consider whether the mine would place the koala population in the area at risk of extinction, as required under threatened species provisions of the *EP&A Act* (s5A(2)(a)). Koala populations in NSW are considered vulnerable under the *Threatened Species Conservation Act 1995* (NSW). Of particular concern was the plan of action for the koala habitat proposed by the proponent, which anticipated that koalas would 'naturally move away from the habitat that is being cleared' and, if they did not, then a translocation plan would be implemented. Evidence was put to the PAC that translocation programs have resulted in high mortality rates for koalas, and Upper Mooki Landcare argued that the PAC did not consider this, constituting an error of law. However, the threshold for a claimant to demonstrate an error of law in judicial review proceedings is particularly high, and Upper Mooki Landcare could not demonstrate that the PAC had failed outright to consider the impacts upon the koala population, or the principles of ESD. Accordingly, Chief Justice Preston dismissed the claim.

Landholders also urged the federal government to intervene under the 'water trigger' (see Box 4.8) of the *Environment Protection and Biodiversity Conservation Act 1999* (Cth) ('*EPBC Act*'). In February 2015, the Federal Environment Minister, Greg Hunt, referred the Shenhua Watermark project to the Independent Expert Scientific Committee on Coal Seam Gas and Large Coal Mining Development (IESC) under the 'water trigger' of the *EPBC Act*. Effectively, the referral 'stopped the clock' on the project, enabling further review of the water-related concerns with the project before a final decision on the project.

BOX 4.8   *EPBC Act* 'water trigger'

The 'water trigger' is one of the 'matters of national environmental significance' under the *EPBC Act* which provides the Commonwealth government with the ability to step in and either approve, refuse or place conditions on a large coal or CSG development where the development will, or is likely to, have a significant impact on a water resource (*EPBC Act* s24D). Other matters of national environmental significance which trigger Commonwealth government involvement include: impacts on World Heritage areas, National Heritage places, Ramsar Wetlands, threatened species, listed migratory species, marine environment and the Great Barrier Reef Marine Park; and the protection of the environment from nuclear actions (*EPBC Act* ss 12–24E).

The 'water trigger' was introduced by the Gillard Labor Commonwealth government in 2013 as a response to widespread community concern about the water-related impacts of coal and CSG developments (Hepburn 2013). The Commonwealth government also established the IESC as a part of the reforms to provide scientific advice to decision makers on the impact of large coal and CSG developments on water resources.

The IESC focused on the potential impacts to groundwater in its assessment, and considered the proponent's revised modelling to be 'sufficiently robust' and accurate in its assessment of likely impacts. However, the IESC expressed doubts as to how groundwater-dependent ecosystems would be affected, as well as expressing uncertainties concerning aquifer impacts at the local scale (IESC 2015). The IESC were also dissatisfied with the proponent's monitoring program and recommended additional monitoring as well as further investigation into aquifer connectivity.

Despite these concerns, in July 2015 the Federal Environment Minister, Greg Hunt, approved the mine, with conditions that included requirements to prepare a water management plan for the Minister's approval, and to stop work if agricultural water supplies are impacted (Department of the Environment 2015a). The decision sparked widespread criticism, with local farmers describing it as 'agricultural genocide' (Hannam 2015b). Minister Hunt claimed that he could not 'have reached a different decision' based on the information before him, and that a decision to reject the mine would have been subject to legal challenge (Akerman 2015). He maintained, in the media, that he had no discretion to weigh up the costs and benefits of the project under the parameters of the *EPBC Act*, and could only consider the advice before him. This frame, however, was arguably inconsistent with the provisions of the *EPBC Act* which provide the Minister with a wide discretion under s 136 to consider 'economic and social matters' in deciding whether or not to approve an action, leading to speculation that this could amount to a legal error that would give rise to judicial review (McGrath 2015). However, the Minister's subsequent statement of reasons indicated that such discretionary matters were considered in his decision (Department of the Environment 2015b), and with the merits of his decision not subject to challenge, the prospects of a successful judicial review were slim.

Public opposition to the mine intensified, gaining significant coverage in the media. During a heated interview on national talkback radio, the Federal Minister of the Environment promised to go beyond his formal obligations to seek further IESC review of Shenhua's water management plan once it was submitted (Winestock and Riordan 2015). However, the Minister subsequently reiterated that the approval had always been conditional upon his acceptance of the proponent's water management plan (Winestock and Riordan 2015). Federal assessment of the project was thus complete, and the only approvals remaining were the grant of a mining lease under the *Mining Act 1992* (NSW), and the pending IESC review of the proponent's water management plan.

The future for coal on the Liverpool Plains?

With Shenhua's exploration licence due to expire in early 2016, and an application for a mining lease still to be lodged, speculation began to mount that the future of the Watermark project was uncertain. In February 2016, Shenhua applied to extend its exploration licence rather than seek a full mining lease, which further fuelled doubts about the project. Meanwhile, the community had expected BHP Billiton to submit a full application and EIS for the Caroona Coal Project sometime in 2015, but by mid-2016 the company had still not done so.

In August 2016, the Premier announced that a landmark agreement had been reached between the state government and BHP Billiton to buy back the Caroona exploration licence for $220 million dollars, $120 million dollars more than what the miner had originally paid (Hannam 2016a). In announcing the buy-back, Premier Mike Baird stated that coal mining under the Liverpool Plains 'poses too great a risk for the future of this food bowl and the underground water sources that support it' (Hannam 2016a). Some were sceptical that the buy-back was prompted by risks to agricultural production and pointed to the now-reduced price of coal as the real rationale; these sceptics criticised the use of public funds to compensate 'a commercial player for a failed investment'. Media reports indicated that the buy-back was motivated by the fierce community opposition to the Shenhua application (Hannam 2016a). The Caroona buy-back has since cast further doubts about the viability of Shenhua's Watermark Coal Project, and recent indications are that the company is reducing its investment in fossil fuel projects (Shenhua 2016). The latest reports indicate that Shenhua has since commenced negotiations with the government to buy back part of its exploration licence (Foley 2016).

In September 2016, the federal government ordered an investigation into the assessment of Indigenous cultural heritage sites in Minister Hunt's conditional approval granted in 2015 (Razaghi, Fuller and Thomas 2016). An application was lodged by the Gomeroi Traditional Custodians under s 10 of the *Aboriginal and Torres Strait Islander Protection Act*, seeking protection of grinding grooves complexes, culturally modified trees, Watermark Mountain and several landscape viewing corridors. The traditional owners maintain that these sites 'retain immeasurable historic, social and cultural value to Gomeroi People', and that Shenhua's mine would destroy the artefacts and disrupt their connection to country (Razaghi et al. 2016). An independent investigation of the claim was conducted in early November 2016, with an open public meeting held to hear oral submissions. Gomeroi elder, Uncle Neville Sampson, said that the sites were sacred and could not be removed, as Shenhua had originally proposed; while Gomeroi Traditional Custodian, Dolly Talbot, noted that the sites were 'a teaching tool for our children and . . . a reminder of the footprints that our ancestors have taken' (Razaghi 2016). A report on the investigation is to be submitted to the Federal Environment Minister by the end of 2016, who will then make a final determination about whether protection of cultural heritage is required.

In October 2016, eight years had passed since the original exploration licence was granted. A provision in that licence allows the state government to cancel the licence if no substantial development of the mine has occurred. Local groups against the mine, including Lock The Gate, Upper Mooki Landcare and SOS Liverpool Plains, wrote to the NSW government to request that the licence be cancelled. However, media reports indicate that Shenhua has sought an exemption to the clause and that the government is considering Shenhua's request (Aston 2016). Members of the community remain strongly opposed to the Shenhua development and have vowed to continue their campaign against its approval.

Conclusion

This first part of the Namoi catchment case study has demonstrated that the experience of injustice is not limited solely to the outcomes of environmental decisions. From a distributive justice standpoint, the impacted landholders and community perceived injustice throughout the earlier phases of the project's appraisal and evaluation – even though the eventual outcome of BHP's Caroona project appears to be favourable to them. The final outcome of Shenhua's Watermark project is still to be determined and landholders and the community point to evidence of injustice throughout the various processes of project assessment as well as in the reform of the overarching land use decision making framework applicable to the project's assessment.

Procedural injustice was a recurrent theme throughout Part 1 of the Namoi catchment case study. From the outset, when coal exploration was first announced in the region, landholders felt that the processes for assessing and approving exploration were unfair. The lack of consultation over the granting of exploration licences, the minimal information available to landholders and other stakeholders concerning exploration activities, and the lack of opportunities to participate in the decision making process for exploration fell short of what landholders expected. Even though, in theory, landholders could challenge land access arbitration decisions, the inability for landholders to be legally represented in arbitration matters, and the narrow parameters upon which they could contest arbitration decisions, left little opportunity to question decisions about land access. Their concerns were not able to be adequately expressed within the limited opportunities for participation at the exploration phase.

Concerns about procedural fairness also extended into the development assessment phase and, as in the Bulga case in Chapter 3, the difficulties associated with obtaining and responding to complex information about proposed development – necessary in order to participate in the decision making process – placed landholders and the community at a distinct disadvantage. The inability to access independent and impartial review bodies to challenge the merits of decisions, such as those made by the PAC and the Federal Environment Minister in the Shenhua case, was a further limitation. The burdens associated with participation created additional negative spillovers, with the economic and emotional transaction costs of participation a hidden sacrifice for many individuals and communities. As Charlie summarised:

> I think the farmers have had economic loss [and] I think the communities have suffered the loss of agriculture . . . [but] the losses are not just financial, they're loss of heart, loss of spirit, loss of enthusiasm, loss of people in the community that actually stay here and farm communities.

These concerns about procedural fairness were inextricably linked to issues of recognition; in several ways, the interests and concerns of those opposed to development were also rendered invisible within the land use decision making process. The failure to account for the special circumstances of the Gomeroi people, which made their participation in decision making processes difficult, is an example of misrecognition reinforced by the imposition of dominant cultural standards in the overarching assessment processes. Another example is in the development of the SRLUP, in which the position of the government shifted little in spite of extensive consultation with the provision of considerable information by communities. Feedback from landholders and community groups was perceived to have been ignored, and the decision making structure was not considered to have been transformed in any meaningful way. Effectively, the land use decision making process became 'colonised' (Groves 2015) by those in power, in which values that could not be understood within the parameters of Strategic Agricultural Land were obscured, further disempowering landholders and the community and frustrating those who had participated in consultative processes in good faith.

Similarly, the preference for expert (generally quantitative) scientific data to measure impact in development assessment processes limited what could be counted as evidence, and required landholders and the community to engage with these data in order to participate effectively in decision making. When objectors did source such data in the Shenhua case, expending a significant amount of money to review the proponent's groundwater modelling, they were ultimately denied the opportunity to challenge the merits of the PAC's evaluation and weighting of their data compared to that of the proponent. The bureaucratic decision to hold a public hearing in the initial Shenhua PAC review closed off future rights to seek merits review in the Land and Environment Court, and effectively removed landholder and community concerns from the realm of formal legal contestation.

Much like the community in the Bulga case, the capabilities of the landholders and others opposed to coal development on the Liverpool Plains were restricted. In particular, the capability of landholders to participate in land use decision making processes – including the formulation of decision making processes – were limited, as were landholder and community capabilities to determine their future – both that of the physical environment and their livelihoods. Following Shenhua's most recent application to extend the term of its exploration licence, one local farmer was quoted in the media as saying:

> We're over it. It's been eight years, they are not doing anything and what gives them the right to hang around leaving everyone else in limbo while other major agricultural projects don't go ahead because people don't know if there's going to be a mine here?
>
> (Aston 2016)

The lack of agency to control such factors points overwhelmingly towards environmental injustice, stemming from an imbalance of power. Interviewees considered the influence of industry as central in this regard. In particular, interviewees pointed to the sizeable exploration licence fees paid by mining companies which created strong expectations that developments would ultimately be approved:

> I think that there's a real problem charging that much for exploration, and it does give rise to a kind of deal done, it's just a matter of getting your ducks lined up and you'll be right . . . you create that expectation that they're likely to be able to exploit that resource.
>
> (Cathleen)

From the outset, the proponents – and the mining industry generally – framed coal exploration and development on the Liverpool Plains as critical to the state's economic prosperity. The government reiterated this frame, particularly following their reform of the *Mining Act* in response to the Supreme Court's decision to overturn certain land access arbitration agreements in the Caroona case. In this instance, the government had power over the governance process and, by its authority, could choose to legitimise particular actors and normalise certain ideas; while the wider industry – and the proponents, more specifically, through their considerable expenditure on licence fees – had power in their interactions with government to encourage the adoption of their preferred scale (Van Lieshout, Dewulf, Aarts and Termeer 2014).

Nevertheless, the discussion of this case demonstrates that relationships of power are not static; that is, the broader economic scale frame did not prevail absolutely. Landholders and their supporters (including state and national farming bodies) vociferously challenged the industry's broader 'economic benefit' frame with counter-frames around the importance of agriculture, not only to the regional and national economy but also to domestic and international food security. These counter-frames enjoyed a groundswell of public support, and – unlike the community in the Bulga case – those opposed to coal development on the Liverpool Plains were also able to exercise power in interactions with the government to motivate reform of the planning system. Power, as noted in Chapter 2, can thus move between actors, particularly in contests over scale.

However, this power did not materialise into a concrete ability to shape the design and implementation of policy; that is, those opposed to coal development did not have the *power of influence* over the land use decision making process (Van Lieshout et al. 2014). The rhetoric surrounding the SRLUP reforms, for example, saw the government's narrative morph into a discourse of 'balancing' industry and agricultural interests. This 're-framing' enabled the government to position itself as unbiased and the new legal framework as an impartial arbiter of land use conflict. In reality, landholder and community concerns were 'filtered from the process' because they did not 'fit the ideas of some decisive actors' (Van Lieshout et al. 2014: 20). As Charlie argued:

> I think the governments stand behind laws . . . as being upholders of the national interest. But when you actually see what they're doing behind the scenes . . . it's nothing more than lip service.

Consequently, the creation and implementation of decision making processes relevant to coal development on the Liverpool Plains may be characterised as a 'power play with different equilibriums' (Van Lieshout et al. 2014: 20). While deliberative mechanisms were promised, they were ultimately driven by the government's *power over* the process, and the actors most able to influence them. This reiterates Lynch's observation, noted in Chapter 2, that an unequal distribution of power in society means that not all environmental perspectives will be considered equally; as such, questions of environmental justice must be concerned not only with the outcomes of environmental decisions but also with whether non-dominant perspectives are 'heard and valued' in decision making processes (1993: 110). Part 1 of the Namoi catchment case study provides another example of how local concerns can be downplayed or even ignored by those in power, not only in the conduct of decision making but also in the ways in which decision making processes are formulated at the outset.

In Part 2 of the Namoi catchment case study (Chapter 5), these issues will be explored further against the backdrop of CSG exploration. While there are some key contextual differences – notably, the different resource proposed for extraction, as well as the location of most the exploration activities on public State Forest land – there are many similarities in terms of how the various elements of environmental justice are perceived, and the role of power in producing injustice.

Notes

1 See e.g. *Mining Act 1992* (NSW); *Petroleum (Onshore) Act 1991* (NSW); *Mineral Resources (Sustainable Development) Act 1990* (Vic); *Mineral Resources Act 1989* (Qld); *Petroleum and Gas (Production and Safety) Act 2004* (Qld); *Mining Act 1971* (SA); *Mining Act 1978* (WA).

2 A copy of the licence is available at: http://ccag.org.au/wp-content/uploads/2013/03/Exploration-Licence-BHP.pdf

3 *Mining Act 1992* (NSW) Sch. 1 cl. 26.

4 Many landholders in the area placed 'Lock The Gate' signs on their properties that withdrew permission for mining companies to enter their property (see www.lockthe-gate.org.au/lock_your_gate). The signs proclaim that a trespass has occurred unless entry is invited by the landholder, citing as authority the High Court case *Plenty v Dillon* (1991) 171 CLR 635 (which confirms that landholders may withdraw an implied right that others have to enter upon their land). Interestingly, this assertion is inconsistent with the provisions of the *Mining Act* (which expressly modifies the common law position noted in *Plenty v Dillon*), but nonetheless proved politically successful with the proponent maintaining publicly that it would not enforce any legal right of entry against unwilling landholders (BHP Billiton 2010).

5 *Margaret Alice Alcorn and Leslie James Alcorn v Coal Mines Australia Pty Ltd* (Case No 2008/57); *Thomas Bailey v Coal Mines Australia Pty Ltd* (Case No 2008/58); *Geoffrey Brown and Sharon Brown v Coal Mines Australia Pty Ltd* (Case No 2008/59); *Anthony Mark Clift v Coal Mines Australia Pty Ltd* (Case No 2008/60), Warden Bailey, delivered 21 May 2009, Mining Warden's Court of New South Wales. Note that appeals against mine access arbitration matters are now heard in the Land and Environment Court.

6 Above case at [11], 6.

7 Above case at [14], 7.

8 *Brown & Anor v Coal Mines Australia; Alcorn & Anor v Coal Mines Australia Pty Ltd* [2010] NSWSC 143.
9 Above case at [96].
10 See www.mpgp.nsw.gov.au/ for more information about the Gateway Panel, and its current membership.
11 Because the Environmental Assessment Requirements were issued in April 2012, prior to the release of the SRLUP in September 2012: *Environmental Planning and Assessment Regulation 2000* (NSW) Cl 50A(3).
12 Because a public hearing was held during the initial PAC review, merits review in the Land and Environment Court was precluded – even though a public meeting had taken place as a part of the subsequent PAC determination process: *EP&A Act* s23F.
13 In particular, provisions relating to Water Sharing Plans under the *Water Management Act 2000* (NSW).
14 A similar issue arose when BHP Billiton submitted its Caroona project plan for the purposes of Gateway Panel assessment. Community members argued that a 'dictionary' definition of floodplain saw floodplain areas included in the longwall underground mine proposal (Foley 2014a). BHP Billiton similarly maintained that their proposal conformed to the definition of 'floodplain' provided to them by government.
15 *Upper Mooki Landcare Inc v Shenhua Watermark Coal Pty Ltd and Minister for Planning* [2016] NSWLEC 6.

5 The Namoi catchment case study: Part 2
Coal seam gas exploration in the Narrabri Shire

Introduction

In the Narrabri Shire, located 160 kilometres north of the Liverpool Plains (see Figure 4.1), cotton and cereal crops abound on fertile black soil plains. Along with numerous other agricultural industries across the area, these crops contribute to the Narrabri Shire's ranking as the second richest agricultural shire in Australia. The gross agricultural production value of the area is A$260 million (ABS 2011b), accounting for approximately 20 per cent of the gross regional product of the Shire (Narrabri Shire Council 2012).

To the south of the agricultural plains lies the State Forests of the Pilliga, eastern Australia's largest intact semi-arid native woodland. Spanning over half a million hectares, The Pilliga (as it is more commonly known) has high species diversity, and is home to nationally and state listed threatened species such as the glossy black cockatoo, brown treecreeper, rufous bettong, koala, squirrel glider and the Pilliga mouse (NSW National Parks and Wildlife Service n.d.). Immortalised by Australian author Eric Rolls, The Pilliga 'is busy with trees, with animals and with men. It is lonely and beautiful. It is a million wild acres. And there is no other forest like it' (Rolls 1981: 1). Underneath the surface of The Pilliga lies the basement rock Pilliga Sandstone formation, which acts as a groundwater recharge for the Great Artesian Basin (see Box 5.1).

BOX 5.1 The Great Artesian Basin

Spanning over 22 per cent of the Australian continent, the Great Artesian Basin lies under parts of the Murray–Darling Basin, and is the largest and deepest artesian basin in the world. It contains some 65,000 cubic kilometres of groundwater that is released from natural springs and artesian bores to provide the only reliable source of fresh water to a significant proportion of inland Eastern Australia (SoilFutures 2015; Queensland Department of Natural Resources and Mines 2012). The groundwater held in the basin is estimated to have taken hundreds of thousands of years to reach the current position, with recharge zones acting as a pressure head to keep the water flowing (SoilFutures 2015).

Historically, The Pilliga was occupied by the Gomeroi (also known as the Kamilaroi) people (Norris, Mitchell and Hart 1991) and the area holds great Indigenous cultural significance with a number of important sites and artefacts recorded (Department of Environment and Climate Change 2008). Following European settlement, The Pilliga, with its abundance of cypress and iron bark trees, became a site of active logging during the early twentieth century (Norris et al. 1991). In 2005, the Pilliga Nature Reserve was created under the NSW *National Parks and Wildlife Act 1968*, eventually expanding to an area of 80,627 hectares set aside for nature conservation. Logging in the remaining State Forest areas continues, but the reservation of National Park and State Conservation Areas has seen the local timber industry decline.

More recently, the area has attracted the attention of companies seeking to carry out CSG extraction (see Box 5.2). With mineral and petroleum exploration and production permissible in State Forests, State Conservation Areas and National Parks, The Pilliga has now become central to NSW's foray into unconventional natural gas development. Amidst broad public debate about unconventional gas, Australian oil and gas company Santos acquired exploration rights in The Pilliga in 2011 and has been conducting exploration and appraisal activities ever since (Santos n.d.). The company is proposing the 'Narrabri Gas Project', claiming it will supply up to half of the NSW's natural gas requirements, as well as bring substantial economic benefits to Narrabri and the wider region (Santos 2015a). However, the proposed development has met with considerable resistance. Of particular concern to those protesting against the project is the loss of critical habitat and impacts upon threatened species within The Pilliga, as well as the potential impact upon the groundwater resources of the Great Artesian Basin, with fears that de-watering to extract CSG will reduce pressure in recharge zones. Compounding these concerns is a legacy of poor environmental management of CSG operations in The Pilliga at the hands of Santos's predecessors.

BOX 5.2   Coal seam gas

Australia has vast reserves of unconventional natural gas, including shale gas and CSG. CSG has been produced in Australia since the mid-1990s, mostly in the Surat and Bowen basins in Queensland (CSIRO 2015; Rutovitz, Harris, Kuruppu and Dunstan 2011). In NSW, the Sydney, Gloucester and Gunnedah basins hold rich sources of CSG.

Largely comprised of methane, CSG is formed in coal seams, where it is adsorbed to the surface of the coal and held under pressure by groundwater. Unlike conventional natural gas extraction, which is released under its own pressure, the extraction of CSG involves reducing the water pressure within the coal seam to allow the gas to 'de-sorb' and be released (CSIRO 2015; WorleyParsons 2013). This is typically achieved by drilling into the coal seam to pump out water ('de-watering'), which transmits water and gas through the well to the surface. Where seams of lower permeability are encountered and the flow of gas is insufficient, fracture simulation (hydraulic fracturing, or 'fracking') - where a combination of sand, water and chemicals are injected into the well to increase the flow of gas - may be required (WorleyParsons 2013).

Until recently, there was little public awareness of the nature and scope of CSG exploration taking place in Australia. For the most part, hydraulic fracturing had not been common in CSG extraction in Australia (O'Kane 2014a; Rutovitz et al. 2011), but now there is widespread public concern concerning its potential use, particularly in light of negative media reports and documentaries on fracking (especially from the US, where fracking is more frequently utilised in shale gas extraction; see Chapter 6). Threats to water quality and quantity, especially in agricultural areas, have dominated public debate, with fears that CSG extraction will place additional pressure on surface and groundwater resources, and contaminate water sources and soils (Chen and Randall 2013; O'Kane 2014a; Rutovitz et al. 2011). Human health and social concerns have also been raised, including the potential for large-scale CSG developments that fragment landscapes and cause biodiversity loss (Chen and Randall 2013; Rutovitz et al. 2011).

This chapter will first describe the background to CSG exploration in The Pilliga, then focus on the current dispute surrounding Santos's Narrabri Gas Project. The assessment process for CSG exploration – as distinct from production – has been at the centre of the conflict to date. The assessment process provides another angle from which to consider aspects of justice in land use decision making. The balance of this chapter examines perceptions of justice within the framework for assessing CSG exploration, drawing on interview data (see Chapter 4, Table 4.2), media reports, public submissions, legislation (Table 5.1), case law and development application documents.

Table 5.1 Legislation, regulations and SEPPs relevant to Part 2 of the Namoi catchment case study

Petroleum (Onshore) Act 1991 (NSW)	Regulates the search for and mining of petroleum in NSW.
Environment Planning and Assessment Act 1979 (NSW) ('EP&A Act')	Regulates the assessment and approval processes for major developments, such as coal mines, within NSW.
Mining Act 1992 (NSW)	Governs the exploration and production of coal within NSW, and vests ownership of subsurface minerals in the state.
Environment Protection and Biodiversity Conservation Act 1999 (Cth) ('EPBC Act')	Governs the protection of the environment and the conservation of biodiversity at the federal level.

Petroleum (Onshore) Act 1991 (NSW)	Regulates the search for and mining of petroleum in NSW.
State Environmental Planning Policy (Mining, Petroleum Production and Extractive Industries) 2007 (NSW) ('Mining SEPP')	Provides specific criteria for the assessment of mining proposals, including compatibility with nearby land uses, any impact of transport, the efficiency of resource recovery and mine rehabilitation.
State Environmental Planning Policy (State and Regional Development) 2011 (NSW)	Provides detailed definitions of what constitutes State Significant Development for the purpose of determining development applications under the *EP&A Act*.
National Parks and Wildlife Act 1968	Regulates the establishment, preservation and management of national parks, historic sites and certain other areas; and the protection of certain fauna, native plants and Aboriginal objects.
Government Information Public Access Act 2009 (NSW) ('GIPA Act')	Facilitates public access to government information.
Freedom of Information Act (1989) (NSW) ('FOI Act')	Provides a right to request access to documents held by government ministers and agencies in NSW. Access may be denied if the documents are considered to be exempt or if their release is deemed contrary to the public interest.
Forestry Regulation 2012 (NSW)	Provides specific regulations for the control and management of forestry areas within NSW.
Administrative Decisions (Judicial Review) Act 1977 (Cth)	Provides for review on questions of law of certain administrative decisions.
Crimes Act 1900 (NSW)	Codifies the criminal law of NSW.
Law Enforcement (Powers and Responsibilities) Act 2002 (NSW)	Sets out police and other law enforcement officers' powers and responsibilities, and the safeguards applicable in respect of persons being investigated for offences.
Inclosed Lands Protection Act 1901 (NSW)	Regulates the protection of inclosed lands from intrusion and trespass.

Coal seam gas exploration in The Pilliga

Petroleum Exploration Licence 238 (PEL238) (see Box 5.3), spanning some 9,900 square kilometres in The Pilliga, was first granted in 1980 to Great Southern Petroleum and Petroleum Securities (Upstream Petroleum Consulting Services 2000). By the late 1990s, the licence had changed hands several times, and various exploration programs had commenced within PEL238 including sites at Wilga Park, Coonarah, Bibblewindi and Bohena (Upstream Petroleum Consulting Services 2000). The pilot programs – located largely on State Forest land – indicated the potential for significant production value, with industry estimating the gas reserves within PEL238 at 35 trillion cubic feet (Morton 2000).

Site inspections by the National Parks Association of NSW towards the end of 2000 revealed little evidence of ongoing exploratory activity in The Pilliga, with drill rigs inactive and in the process of being dismantled (Western Conservation Alliance et al. 2001). Atkinson (2005) also reported evidence of environmental damage at the Bohena site: a holding dam, which held highly saline wastewater pumped from underground during the extraction process, had collapsed and saline liquid had leaked through to the subsoil and shallow aquifers; a black sludge substance was found on the forest floor (Atkinson 2005; Western Conservation Alliance et al. 2001); there were dead animals discovered nearby; and surrounding vegetation was dying at sites up to one kilometre away (Atkinson 2005). The NSW EPA investigated the site in early 2001 and clean-up work commenced shortly after.

Eastern Star Gas (ESG) acquired PEL238 in 2001 and, by 2004, had renewed exploration activities in The Pilliga with the drilling of several new wells and the construction of the Wilga Park Power Station to commercialise conventional gas extracted from the Coonarah gas field (Atkinson 2005; Eastern Star Gas 2010). Ongoing exploration at the Bibblewindi and Bohena pilot sites confirmed the presence of significant CSG resources. In 2007, the NSW government granted a Petroleum Assessment Lease (PAL2) (see Box 5.3) for the Bohena area within PEL238 to enable the recovery of gas from exploration activities. In 2008, ESG successfully gained major project approval from the Minister for Planning (under former Part 3A of the *EP&A Act*; see Chapter 3, Box 3.2) for the 'Narrabri Coal Seam Gas Utilisation Project', an exploration and appraisal program which saw further infrastructure built, including: a gas gathering system, gas compression facilities, a 32-kilometre underground gas flow line and the expansion of the capacity of the Wilga Park Power Station to utilise the CSG generated from its pilot sites, which would otherwise have been vented and wasted (Department of Planning 2008).

BOX 5.3   Regulation of unconventional gas exploration in NSW

Exploration of unconventional gas (including CSG) in NSW is covered by the *Petroleum (Onshore) Act*, which, like the *Mining Act* in the case of minerals, vests ownership of subsurface petroleum resources in the Crown (s6). There are two main types of exploration titles under the *Act*:

1. Exploration licences, which provide an exclusive right to prospect for petroleum (s29); and
2. Assessment leases, which enable further prospecting and recovery of petroleum in the course of assessing the viability of commercial recover (s33).

By 2010, exploration activities had extended to five separate pilot production fields containing more than 90 wells, along with the construction of dams, water treatment facilities and other infrastructure, such as roads. ESG then sought project approval under former Part 3A of the *EP&A Act* (see Chapter 3, Box 3.2) for the 'Narrabri Coal Seam Gas Project' covering 7,920 square kilometres of State Forest and freehold land within PEL238. The project enabled the drilling of 1,100 wells, along with the construction of gas processing and compression facilities, dedicated water management and treatment facilities, and high pressure gas transmission pipeline infrastructure (Eastern Star Gas 2010). The preliminary environmental assessment acknowledged, however, that there would be several risks associated with the project, including an increased risk of bushfire, reduced habitat for threatened species, loss of native vegetation, potential for contamination of soil and surface water, and depletion of aquifers (Eastern Star Gas 2010).

On several occasions throughout 2011, residents reported, to several authorities, evidence of leaking water storage ponds, spillage of highly saline produced CSG water, leaking gas pipes, dead native animals and vegetation dieback in The Pilliga. However, no responses were ever received to these alleged environmental incidents (Northern Inland Council for the Environment 2012). Soon after, protests commenced in earnest against ESG's exploration activities and their planned Narrabri Coal Seam Gas Project, including 'lock-ons' where protesters attached themselves to drilling equipment to disrupt and delay exploration operations. Community members and environmental groups believed that such direct action was a necessary last ditch effort to express their concerns with the proposed project (Clifford 2011), the extent of which, they believed, had not been fully revealed in the preliminary environmental assessment submitted by ESG (Friends of the Pilliga 2011a). Environmental groups also contended that ESG's exploration activities would have a significant impact on matters of national environmental significance under the Commonwealth *EPBC Act* and urged the Commonwealth government to intervene (Northern Inland Council for the Environment et al. 2011; Jordan, Clarke and Flint 2011), which it did, acknowledging that the activities would have a significant impact on threatened species and could be classified as controlled activities requiring an EIS and development approval.

During the 2011 NSW state government election, public concern about CSG exploration and development had escalated and become a critical issue (along with coal development; see Chapter 4). Central to these concerns was a lack of public confidence in the overarching state regulatory framework which governs CSG exploration and production, which has been described as 'highly complex and fragmented' with 'considerable cross-referencing, layering, exceptions and variations within and between legislative instruments' (O'Kane 2014b: 1). A particular issue was the extent of CSG exploration licences that had been granted throughout the state with limited public consultation. The newly elected O'Farrell Liberal/National Coalition government placed a moratorium on hydraulic fracturing, a ban on the use of hazardous BTEX chemicals (benzene, toluene, ethylbenzene and xylenes) as additives in CSG drilling, updated public consultation guidelines (Hartcher 2011), placed a moratorium on the granting of new CSG exploration licences and introduced the SRLUP (see Chapter 4) (Hazzard 2011). However, environmental and community groups in The Pilliga were critical of these measures – particularly the moratoria on new exploration licences and hydraulic fracturing, which were seen as 'Clayton's moratoriums',[1] because they were not applicable to the already established exploration activities underway in The Pilliga.

In late 2011, Santos (a major shareholder of ESG) completed a successful A$924 million takeover of ESG and withdrew ESG's Part 3A project application (Santos 2011; Cubby and Nicholls 2011). Santos foreshadowed, at the time, that it would continue exploration work in The Pilliga with a view to proposing its own 'Narrabri Gas Project' for review under the new State Significant Development provisions in Part 4 of the *EP&A Act* (Santos 2011).

The Narrabri Gas Project

By the time Santos took over ESG's exploration operations, the community was extremely frustrated with the limited opportunities for consultation over the existing program of exploration, and were concerned about the growing evidence of environmental damage in The Pilliga (Northern Inland Council for the Environment 2012). Interviewees for this research considered that the broad Ministerial discretion to approve ESG's initial Narrabri Coal Seam Gas Utilisation Project as a major project under former Part 3A of the *EP&A Act* (which provided limited opportunities for public consultation), coupled with the lack of public information about the extent of environmental damage, had resulted in a failure to meaningfully engage the public from the earliest possible opportunity – a fundamental plank of procedural fairness (Chapter 2). Santos was, thus, faced with the immediate challenge of repairing the reputation of CSG exploration operations in The Pilliga; a difficult task given its prior association with ESG as a major shareholder.

Initially, Santos blamed excessive rain for water pooling (and the presence of eucalyptus leaves for its discolouration). However, in early 2012, Santos released a report on ESG's operations, admitting that there had been an 'unacceptable culture in Eastern Star of accepting minor spills, failures in reporting and the possibility of unapproved land clearing on some sites' (Santos 2012). The report confirmed many of the fears that had been expressed about CSG operations in The Pilliga. The EPA fined Santos A$3,000 for ESG's discharge of polluted water and issued a formal warning (EPA 2012).

Santos shut down existing operations, including decommissioning an existing water treatment facility, and pledged A$20 million to remediate and upgrade operations before recommencing exploration (Santos 2012). Environmental and community groups remained sceptical about the actions of Santos, noting that there was 'little or no information available publicly' on any of the government investigations of the community's complaints of breaches, and no 'obvious opportunity for public input' into the relevant investigative processes (Northern Inland Council for the Environment 2012: 13). The EPA's eventual penalties were dismissed by one group as 'hitting the gas company with a piece of wet lettuce'; the community had spent more money conducting the necessary tests to provide evidence of contamination than what the company had ultimately been fined (Friends of the Pilliga 2012a).

Despite these setbacks, Santos remained committed to its planned Narrabri Gas Project. In late 2012, Santos announced that it would renew its program of exploration in The Pilliga.

'Kept in the dark': public participation during exploration assessment

Santos conducted a range of outreach activities to communicate with the public about its planned exploration activities and the foreshadowed Narrabri Gas Project, including community open days. They also established a Community Consultative Committee (CCC), which met monthly. However, interviewees for this research felt that these venues did not provide an adequate opportunity for their concerns to be discussed and addressed. There was a perception that the community was only being provided with the information that the company wanted it to hear, and that that crucial information was being withheld. For example, on one occasion, the Wilderness Society attempted to uncover information from 34 documents held by the Department of Trade and Investment, Regional Infrastructure and Services (DTIRIS) about Santos's exploration activities under the *GIPA Act*. However, these requests were denied because Santos objected to the release of the information (Friends of the Pilliga 2012b). As one spokesperson for the Wilderness Society said:

> By objecting to the public release of these coal seam gas documents, Santos is acting as if they have something to hide . . . they took over coal seam gas operations in the Pilliga Forest promising an open approach but instead we are lumped with tactics of secrecy and concealment around the impacts of coal seam gas on our waterways and forests.
>
> (Friends of the Pilliga 2012b)

Interviewees claimed that '[e]verybody was guessing' and that the community was being 'kept in the dark' (Belinda), with 'no transparency' and important information kept 'behind closed doors' (Adam). As highlighted in Chapter 2, provision of information that is effective, timely and representative of public concerns is an important component of procedural fairness. In this case, the community considered the withholding of what was deemed to be critical information as a breach of the community's 'basic right to know what is going on' (Friends of the Pilliga 2012b). Moreover, it evinced a failure to recognise the community as a key stakeholder in the future of The Pilliga. Community groups expressed anger that

Santos's failure to consult them in the preparation of their draft proposal was a rejection of their interests and values:

> Santos should have come to the community first with a draft proposal and conducted a full public exhibition process so that people could have a say in what happened and where. To show the community a map with locations of drill sites identified should be a basic part of Santos consultation. Instead, they seem to have finalised their plan in secret first and now they are going to tell us about it – this is just like a slap in the face for everyone that lives here.
>
> (Friends of the Pilliga 2012c)

Interviewees were similarly critical that opportunities for public consultation and participation would only be triggered once a formal development application and EIS was submitted:

> There is no public participation [at the exploration stage] – what would we participate in? There is nothing to participate in, there is no forum. Who from the government would we talk to?
>
> (Alison)

> There's nobody to tell you what they're doing, nobody's telling us what they're doing . . . All these things are happening and nobody knows why. It'd be nice if they just, sort of, could tell you a little bit more.
>
> (Belinda)

These sentiments reiterate the fundamental link between participation and recognition (Chapter 2). As Schlosberg notes, '[i]f you are not recognized you do not participate; if you do not participate you are not recognized' (2007: 26). Santos's failure to include the community in the early phases of developing their program of exploration, and their suppression of key information, ignored the interests of the community and constrained their ability to participate in the decision making process concerning exploration activities in The Pilliga.

Santos sought and gained approval for the resumption of exploration activities in 2013 under Part 5 of the *EP&A Act* (see Box 5.4), including the drilling of new pilot wells at the existing Dewhurst site within the Pilliga East State Forest, and the construction and operation of the first phase of a water treatment facility (to be located at its Leewood property, adjacent to the State Forest) (Santos 2013a). At that time, while most petroleum exploration activities required development consent as SSD,[2] some exceptions applied; for example, where the development involved five or fewer wells more than three kilometres from another petroleum well,[3] or where the development of related works (such as pipelines and processing plants) is not ancillary to another SSD.[4] In this case, a Part 5 assessment was deemed appropriate for the scope and nature of the work proposed. A Part 5 assessment typically involves the Division of Resources and Energy within the

Department of Industry (previously DTIRIS) considering a proponent's REF, taking into account to the fullest extent possible all matters which are likely to affect the environment (*EP&A Act* s 111). Members of the public may be consulted by the proponent in the preparation of a REF but they do not have the opportunity to challenge a REF submitted in support of a Part 5 assessment (as they would for an EIS submitted for a full development activity under Part 4 of the *EP&A Act*).

BOX 5.4   Assessment of unconventional gas exploration activities in NSW

While some exploration activities, such as the construction of pipelines and processing plants, require full development assessment under the *EP&A Act*, in most cases environmental assessment for exploration activities is typically undertaken under Part 5 of the *EP&A Act*. The most common form of assessment under Part 5 is a departmental Review of Environmental Factors (REF) (see Chapter 4, Box 4.1).

In some cases, where exploration activities will be likely to have a significant impact on the environment (including threatened species, populations and ecological communities), an environmental impact statement (EIS) must also be prepared and placed on public exhibition for at least 30 days (*EP&A Act* ss 112, 113).

Once again, the community expressed a sense of injustice with what it perceived to be inadequate participatory processes, and a lack of recognition of their interests and concerns. With no opportunity for formal public scrutiny, interviewees for this case study were quite critical of the REF as a basis for assessing the impact of Santos's exploration activities. Interviewees believed that there were several inaccuracies in the REF, such as targeted flora and fauna surveys which failed to represent whole-of-site impacts, as well as missing and out-of-date data, but they had no opportunity to challenge it (see also Jordan et al. 2011; and EDO 2011). The nature of public consultation in the formulation of the REF document was also deemed ineffective:

> They have their own appointed little community consultative group where they round up one farmer and one shopkeeper and one bus driver and they're the token tick-box community consultative group.
>
> (Adam)

Overall, interviewees felt that the REF lacked rigour when compared to a full EIS. As Samuel remarked, REFs 'tend to be very crudely done documents', and were seen to be insufficient for making an informed decision about an activity that could potentially have a significant environmental impact. As Alison summarised:

> I think an EIS gives people a bit more comfort. I think why it is difficult is the uncertainty, and that is what the EIS can do, give you much more certainty about the operations, the impact, it just gives you a structure that you can understand . . . It is really hard to weigh it up against the reality without something like an EIS.

In June 2013, Santos announced that it would seek approval for a 'more focused exploration and appraisal program' in The Pilliga, which involved recommissioning existing ESG pilot wells and the construction of new pilot wells to test the viability of the Narrabri Gas Project (Santos 2013b). The proposed exploration activity was deemed a SSD, requiring a full EIS (*EP&A Act* s 78A (8A)). Santos's proposals were contained in three separate applications:

1.   a modification to the existing Narrabri Coal Seam Gas Utilisation Project approval acquired from ESG;
2.   a new proposal for the Bibblewindi Gas Exploration Pilot Expansion; and
3.   a new proposal for the Dewhurst Gas Exploration Pilot Expansion.

Members of the public had an opportunity to review and make formal submissions on the proposals. The majority of the 380 submissions received on the Bibblewindi and Dewhurst applications objected to the exploration activities, expressing concerns relating to contamination and depletion of water resources. The applications were also referred to the Commonwealth government for assessment under the 'water trigger' of the *EPBC Act* (s24D) (see Chapter 4, Box 4.8). A further 3,000 submissions were received, predominantly opposed to the exploration and appraisal proposal. Nevertheless, the Federal Environment Minister determined that the exploration activities did not require Commonwealth approval because they were not of a sufficient scale to have a significant impact on water resources (Department of the Environment 2013).

Many of the submissions against Santos's proposals had expressed concern about the foreshadowed Narrabri Gas Project, which members of the public understood (from Santos's public relations activities) would be a large-scale development comprising some 850 wells. These concerns were again reiterated at a combined PAC public meeting held to consider the Bibblewindi and Dewhurst pilot expansions in June 2014. However, in its determination report, the PAC noted that views about the foreshadowed project were not relevant to the assessment of the present exploration expansion applications, and disregarded these concerns. Overall, the PAC considered the predicted impacts to be 'minor', given the nature and duration of exploration activities, and approved the exploration expansion program in July 2014 (Planning Assessment Commission 2014b).

Community members believed that their inability to express fears about potential future impacts of the full Narrabri Gas Project within the confines of the exploration assessment process overlooked the very essence of their concerns:

> [A]re we still going to have a community in 25, 30 years' time? That's what I think a lot of people are asking . . . maybe they won't damage our aquifers . . . But there's every likely chance that they will and once it's gone, it's gone . . . I think, is the pushing point for the farmers, and for the community members . . . It means everything to them.
>
> (Belinda)

By rendering submissions and statements that related to the complex issue of the region's long-term future as irrelevant, the assessment framework denied the community the opportunity to articulate fundamental place-based concerns. As discussed in Chapter 2, such a lack of recognition not only impairs participation in environmental decision making but can also restrict a community's agency to exercise control over their local environment and their future in general.

'Backed into a corner'

When Santos recommenced exploration activities in The Pilliga, protests in the State Forest intensified. Sit-ins and lock-ons became a regular occurrence; local court lists became congested with civil disobedience charges against protesters (Chillingworth 2014; Clifford 2014), which included local farmers and community members (Foley 2014b; Hasham 2014). The local newspaper quoted one farmer as saying:

> We feel we have been backed into a corner by governments refusing to listen to us or to consider the impacts of the potential loss of our groundwater if this gasfield is developed.
>
> (Namoi Valley Independent 2013)

Interviewees for this research agreed:

> We've exhausted every avenue of writing to ministers and writing to people and nobody's listening . . . Everybody you talk to feels that there's nowhere to go to put forward our [concerns] . . . we're stymied . . . and if we have to lock on, well –
>
> (Belinda)

> Well if you think about what we have to do to get our voices heard . . . it would seem as though the only way that the people are heard is to block stuff.
>
> (Susan)

In the midst of community protests Santos was prosecuted under s 136A(1) of the *Petroleum (Onshore) Act 1991* (NSW) for ESG's failure to report a June 2011 spill of 10,000 litres of produced CSG water from the Bibblewindi water treatment plant and a failure to accurately report on compliance with environmental standards to the Department any incidents causing or threatening material harm to the environment. This was the first time a company had been prosecuted under this section of the *Act*; it resulted in a $52,500 fine against Santos and an order for costs.[5] Shortly after this, the EPA fined Santos a further $1,500 for contaminating a groundwater aquifer through a leaking water storage pond (EPA 2014).

Two days after the EPA issued its fine, the state government signed a Memorandum of Understanding (MOU) with Santos, declaring their proposed Narrabri Gas Project a 'strategic energy project' for NSW because of its potential to supply up to half of the state's natural gas needs (Stoner and Roberts 2014). The then-acting Premier stated that the MOU did not guarantee certainty of outcome and would not bypass development assessment processes, but the government did agree to make a final decision on the project by 23 January 2015. Interviewees for this case study expressed outrage that the project had received what appeared to be favourable treatment in the face of persisting environmental concerns about the project – and with Santos yet to publicly share a comprehensive project plan for the entire Narrabri Gas Project:

> Days before . . . contamination issues were bought up in The Pilliga. The next minute, [the acting Premier] is signing an MOU to fast track the Pilliga project!
>
> (Julia)

> If that's an example of good governance, I don't know what is.
>
> (Samuel)

> For [the acting Premier] to come out and say he's going to fast-track this Narrabri project was just like a slap in the face.
>
> (Belinda)

Santos and the government advanced scale frames in the media to reiterate the energy and economic priorities of the state and the capacity for the Narrabri Gas Project to meet those needs (Santos 2014; McHugh, Condon, and Herbert 2014). In April 2014 the Forestry Corporation of NSW, responsible for State Forests, closed public access to parts of the Pilliga East State Forest under the *Forestry Regulation 2012* (NSW). Santos retained exclusive use of the forest during the closure period. There was a lack of public information about the closure, which led to suspicions about why it had taken place:

> Why shut the forest? Are they hiding something? I don't know . . . to me it doesn't make them look very good, you know, in the eyes of the community.
>
> (Belinda)

The Forestry Corporation maintained publicly that it had been directed by the police to stop access due to public safety concerns (Clifford 2014), but interviewees for this case study felt that it was a politically motivated manoeuvre to stifle legitimate protest and dissent:

> The way they're going they're just trying to deny the average public person input into decision making in general . . . We thought . . . this sort of thing [was in] the past where governments just run roughshod over people and just had their way wherever. I wouldn't be surprised if we see new laws to counter activism and counter protesters.
>
> (Adam)

Indeed, as Adam predicted, this is precisely what happened at both state and federal levels. In 2015, the Commonwealth government sought to repeal s 487 of the *EPBC Act*, which provides a broad standing to conservation groups to bring judicial review proceedings against development approvals under the *Act*.[6] The Bill sought to limit judicial review proceedings to those who are directly impacted by a development (under the common law 'person aggrieved' test: *Administrative Decisions (Judicial Review) Act 1977* (Cth)). The government's rationale for introducing the Bill was the 'emerging risk' that the current laws were being used by 'vigilante green groups' to 'sabotage' and 'deliberately disrupt and delay key projects' (Hasham 2016), despite the fact that litigation under this provision to date has been extremely limited. At the time of writing, however, the proposed legislation had lapsed in the Senate.

In early 2016, the NSW state government successfully introduced new police powers and penalties for trespassing on mining and CSG operations and lock-ons to equipment. In particular, specific offences were added to the *Crimes Act 1900* (NSW) for 'interfering with a mine' (including CSG operations). Intentionally or recklessly destroying or damaging equipment, buildings or roads associated with a mine, or hindering the working of equipment belonging to or associated with a mine can now attract a penalty of up to seven years imprisonment (s 201). The *Law Enforcement (Powers and Responsibilities) Act 2002* (NSW) was also expanded to enable police to stop, search and detain individuals where there is a suspicion on reasonable grounds that a person has, in their possession or under their control, anything that is intended to be used in a lock-on and is likely to be used in a manner that will give rise to a serious risk to their safety (s 45C). A new offence of 'aggravated unlawful entry on inclosed lands' was also created under the *Inclosed Lands Protection Act 1901* (NSW), with a maximum penalty of A$5,500 for trespass and interference with the conduct of business on inclosed lands.

Both the Commonwealth government's attempt to limit standing under the *EPBC* and the state government's expansion of police powers and penalties concerning protest activity constituted a further denial of the interests of the broader community in the Pilliga development. By characterising those who challenge developments as 'vigilantes' using 'green lawfare', the

Commonwealth government used derogatory terms to denigrate and disrespect the interests and social status of those opposed to extractive development (see also Chapter 2). While the Commonwealth government's efforts at reform were ultimately not realised, the pejorative labelling of those who exercised legitimate rights of participation was a clear example of malrecognition. Suggesting that those who opposed development had engaged in 'lawfare' cast their interests as illegitimate and outside the boundaries of the law, and their attempts to engage the legal system in this way was characterised as an abuse of process. Narrowly confining the concept of 'legitimate interest' to those individuals or corporations with an economic interest at stake, was, as Clark notes, 'out of step with emerging international norms around public participation [and] the right of the community to participate in decisions that affect the environment' (2016: 258).

At the state level, the asymmetry between the penalties faced by protestors, and the fines previously imposed upon Santos for environmental breaches, compounded misrecognition further. Ultimately, the provisions have since discouraged protest activity in the area and have closed off what many community members saw as the only way to bring attention to their concerns. Many have been critical that these reforms are an attack on freedom of association and democratic rights of protest, and were 'clearly intended to shore up Santos and the gas industry' (Nicholls and Hannam 2016). Indeed, documents since obtained by the Wilderness Society under the *FOI Act* indicate that Santos had communicated with departmental advisors regarding the new laws before they were read in parliament, raising suspicions of industry influence in amending the law (Hannam 2016b).

Incremental approvals

In 2014, with its expanded exploration and appraisal program approved and underway, Santos had lodged a Preliminary Environmental Assessment for the full Narrabri Gas Project to the Department of Planning and Infrastructure in order to obtain the SEAR necessary to prepare an EIS. The preliminary assessment indicated that the project would involve:

- further exploration and appraisal activities (including ten sets of four-well pilots);
- installation and operation of up to 850 individual production wells;
- a gas processing facility;
- water management and treatment facilities; and
- other supporting infrastructure.

Interviewees for this research indicated that the submission of an EIS would provide them with an opportunity to formally review and respond to Santos's entire Narrabri Gas Project. However, the community also foreshadowed the constraints they would experience in participating effectively within limited public exhibition time frames:

> We're going to get our 30 days, but we're community members, you know, we're going to be landed with a document that's probably 4,000 pages, which is what's happened in Queensland, where the farmers have been presented with a document that's 4,000 pages, now they've got to respond in 30 days. Impossible. That's social injustice, I think.
>
> (Belinda)

> I think from a local council perspective . . . it does come down to capacity and even [the capacity of] local residents. You know, from my experience, local residents don't feel confident to participate because they feel like it is way over their heads, and what would they have to contribute and 'how would I know?'
>
> (Robert)

The EIS was expected to be submitted during 2015 but, as of mid-2016, it is still to be lodged. In the interim, Santos sought approval for Phase 2 of its Leewood water treatment facility under Part 5 of the *EP&A Act*. The proposal involves the construction and operation of a reverse osmosis plant and brine concentrator, and other associated infrastructure for the processing and treatment of wastewater stored in the Leewood ponds. Santos considered these activities to be part of its exploration and appraisal program, and permissible without Part 4 development consent (Santos 2015b). Santos prepared a REF, sent it to various government departments and selected community and environmental groups for comment.

Several concerns were raised about the Leewood proposal, including the lack of data about the volume and quantity of produced water and brine concentrate, lack of information about the disposal of concentrated brine, and lack of analysis of the impact to groundwater sources. With the project located in an area that is an important recharge zone for the Great Artesian Basin (SoilFutures 2015; Smerdon, Ransley, Radke and Kellett 2012), there was a significant concern that the highly saline produced water would affect the quality of the water in the Basin over time. In light of these issues, some of the respondents indicated that it would be more appropriate for the proposal to undergo full development assessment to enable independent review of Santos's claims. Santos, however, considered that the appropriate process had been followed and deemed many of the questions raised about the Leewood REF as 'irrelevant' to the proposal submitted (Sturmer and Armitage 2015). The Department of Industry agreed, and approved the facility.

Interviewees for this case study believed that Santos was proposing development activities incrementally, using Part 5 of the *EP&A Act* to avoid scrutiny of the cumulative impact of the project. As Cathleen noted:

> I think CSG has entered the spectrum through the traditional idea of 'exploration is a relatively benign activity on the landscape so we don't need to

worry too much'... Personally, I think that exploration in terms of coal seam gas projects and exploration in terms of [coal] mining projects – we need to finally accept that they are very, very different creatures. We're still kind of looking at them through the same lens – its exploration, its pre-production ... but the example we've got in The Pilliga, the evidence is just overwhelming. You just cannot compare the two.

In 2016, the community group, People for the Plains, challenged the Leewood approval, seeking an injunction in the NSW Land and Environment Court to prevent Santos from developing the Leewood wastewater treatment facility on the grounds that it had been improperly assessed. Specifically, People for the Plains argued that the process of approving the project as an extension of exploration activities was invalid because the Leewood project was more appropriately characterised as a waste management facility[7] that should have been assessed separately with its own EIS and subject to public exhibition and review. However, the Land and Environment Court disagreed and dismissed the appeal.[8] The court's decision confirms that exploration title holders may conduct activities that might otherwise require development consent where those activities 'serve the purpose' of exploration. An appeal has since been lodged in the NSW Court of Appeal to challenge the decision but, at the time of writing, a decision had not yet been rendered.

The future for coal seam gas in The Pilliga?

In 2014, the NSW Chief Scientist and Engineer, Professor Mary O'Kane, completed an independent review of CSG activities in NSW (O'Kane 2014a). Based on 19 months of research into the CSG industry, including the overarching regulatory framework for CSG operations, the report acknowledged that there were risks associated with CSG exploration and production. Nonetheless, it was considered that, with effective consultation mechanisms and a 'world class' regulatory regime, those risks could be appropriately managed.

The NSW government has since responded to this review with the release of its 'Gas Plan', which sets out a number of reforms to the regulatory arrangements for CSG exploration and production in the state, including:

- the appointment of the EPA as the lead regulator for gas exploration and production;
- changes to the assessment of exploration activities;
- extinguishing current PEL applications and offering 'buy-back' opportunities for current PEL holders;
- amending current PELs to remove exploration areas granted over National Parks;
- the introduction of a Strategic Release Framework, which will limit development to approximately 15 per cent of the state and subject to tender;

- a 'use it or lose it' policy, which will see titles extinguished where a proponent fails to engage in exploration or production; and
- changes to provisions regarding land access and arbitration (NSW Government 2014).

In addition to changes already made, including the SRLUP, Gateway process, Aquifer Interference Policy, and codes of practice on well integrity and fracture simulation, the state government believes it now has the balance right (NSW Government 2014).

At the time of writing, some of the proposed amendments under the Gas Plan have been implemented, including the cancellation of existing PEL applications to reset the approach to issuing exploration licences. The Strategic Release Framework policy document has also been drafted and is the subject to public consultation.

Changes to the assessment of CSG exploration activities have also been introduced through the removal of the 'five wells' rule that required development consent for exploration activities involving more than five wells within a three-kilometre radius. Under amendments to the *Mining SEPP*, petroleum exploration is now clearly listed as development permissible without consent, as is the construction, maintenance or use of pollution control works and equipment associated with petroleum production (Reg 6). As a result of these changes, the Minister for Resources and Energy will now determine all exploration activity under Part 5 of the *EP&A Act* following the assessment of a REF.[9] An EIS must be prepared in situations in which the proposed activity is considered to have a significant impact on the environment. Further amendments to the *Petroleum (Onshore) Act 1991* (NSW) also confer rights on exploration licence holders to beneficially use gas recovered in the process of petroleum exploration (for example, using it or selling it) (s 28B), such as for irrigation of crops (as proposed for Santos's Leewood facility).

The impact of these legislative and policy changes remains to be seen. In the meantime, the community awaits the submission of a full EIS for Santos's Narrabri Gas Project and to have an opportunity to formally review the full proposal. Both Santos and the state government have continued to frame NSW as facing a shortage of gas supply in the near future, which will place jobs in gas-dependent industries in jeopardy and force the state to compete with export markets for interstate gas supplies unless CSG production in NSW can move ahead quickly (Holloway 2015; Godfrey 2016). Indeed, the chief executive of Santos has maintained that the Narrabri Gas Project is 'a highly strategic asset, because NSW is going to be short of gas', adding that '[n]obody should doubt our resolve to bring this to market' (Chambers 2015). However, the Australian Energy Market Operator's most recent *Gas Statement of Opportunities* indicates that there is no short-term risk to gas supply in NSW. It states that declining consumption and upgrades to infrastructure capacity mean that existing supply sources (largely from interstate) are still capable of meeting demand without the need to rush to develop NSW's CSG reserves, including those in The Pilliga (Australian Energy Market Operator 2015).

Conclusion

This second part of the Namoi catchment case study has further demonstrated that injustice may be perceived in decision making processes as much as in the eventual outcomes. In particular, it shows that injustice can be experienced even before formal decision making processes commence where a lack of consultation and information about upcoming environmental decisions fails to adequately engage with community concerns. Such experiences of injustice align with one of the central tenets of the concept of environmental justice: that fair treatment and meaningful inclusion in decision making processes can be equally as important as the outcomes of those processes.

From an environmental justice perspective, procedural inequity was evident even before Santos took over exploration operations in The Pilliga. The public had very limited opportunities to participate in the decision making processes surrounding ESG's exploration activities, and had an ongoing concern that the environmental impacts of exploration were not being well managed. After taking over ESG's operations, Santos faced the difficult task of building trust in its operations, yet it too failed to meet expectations in terms of procedural fairness; stakeholders considered the public open days and community consultation events held by Santos to be superficial and not representative of the particular concerns held by the public. Santos's subsequent objection to the release of certain documents when requested by the public further heightened the sense of procedural injustice.

The public also considered it unfair that certain exploration activities could be approved under Part 5 of the *EP&A Act*, and believed that the use of a REF to assess proposed activities lacked necessary rigour and was not open to challenge by members of the public. Ultimately, the community's concern was that such tactics would enable the company to incrementally secure approval for separate components of development by avoiding cumulative impact assessment of the whole project of works.

Once again, as with Part 1 of the Namoi catchment case study and the Bulga case study, concerns about procedural fairness were entwined with issues of recognition. For example, Santos's rejection of requests for documents under the GIPA Act was not only a failure to provide satisfactory information but also a failure to recognise and respond to the community's concerns. Further, the processes for assessing exploration activities, which were limited to the impacts of exploration only, denied those opposed to development the opportunity to express longer-term place-based interests. The foreshadowed Narrabri Gas Project raised significant uncertainties about future impacts to water resources in the region yet these were deemed irrelevant to the assessment of short-term exploration activities. This framing restricted the capacity of landholders and the community to raise more complex concerns about place transformation; to 'ensure that the natural resources that we have here are looked after, so that we can sustain ourselves indefinitely' (John). Such sentiments could not be captured within the limited framework for assessing exploration activities.

The restriction of public access to the forest, along with the implementation of new penalties and police powers with respect to protest activity, closed off what many perceived to be their only opportunity to have their voices heard. Effectively, the institutional arrangements acted to prevent the emergence of conflict rather than attempting to manage the conflict in a constructive way (Martin and Kennedy 2016). The MOU reached between Santos and the state government to establish the Narrabri Gas Project as a strategic energy project for the state was an example of nonrecognition, which, when combined with various acts of scale framing, positioned the Narrabri Gas Project as essential to the state's economic and energy requirements and obscured the concerns of those who objected to development. Attempts at the federal level to limit standing for judicial review of approvals, and the inflammatory and derogatory labelling of those who sought to challenge development in general further impaired the recognition of those opposed to CSG exploration in The Pilliga.

These points about recognition highlight an interesting dynamic central to the second part of the Namoi catchment case study: the location of the Narrabri Gas Project, predominantly on public land, and who, then, had a legitimate interest in the project. Nearby landholders believed they had an interest because of the project's direct impacts upon biodiversity, water resources and agricultural activities; others, including non-local individuals and groups, believed they had a legitimate interest because of the broader concerns with climate change and environmental protection. As noted in chapters 1 and 2, human capabilities depend upon a functioning environment, and there has been a growing international recognition of the rights of nature as well as the rights of the public to a clean and healthy environment. However, such rights have not been recognised in Australia, despite increasing community expectations of environmental protection and participation in environmental decision making (Clark 2016). Part 2 of the Namoi catchment case study emphasises the importance of a broad interpretation of standing and access to independent review functions to ensure that all interests in environmental decisions are accounted for.

Even though Santos's Narrabri Gas Project is still at the exploratory phase, the capabilities of those opposed to development were restricted. There was little opportunity to participate in decision making regarding CSG exploration, and the inability to articulate what was considered important – protection of the environment, and preservation of place – restricted the freedom of those opposed to exploration to deal with threatened place disruption. As with Part 1 of the Namoi catchment case study, interviewees attributed this lack of agency to the overwhelming power of the government and the political influence of the proponent. The government had power over decision making processes while Santos had power in its interactions with the government – through its expenditure on exploration licence fees and promised royalty revenue – to influence particular outcomes. As Robert remarked, 'in terms of the local community wielding any political power, we don't really'.

Existing extractive development in the Namoi catchment is already substantial, but when overlaid with planned coal and CSG projects from the Liverpool Plains and The Pilliga the cumulative impacts will be undeniably significant. Parts 1 and 2 of the Namoi catchment case study have provided a glimpse of a region attempting to come to terms with the potential impacts and benefits of increased extractive development and their capabilities to direct whether and how development should proceed. The case studies show that both the existing regulatory structure and the subsequent reforms to it have neglected to respond to the fundamental concerns of those who were opposed to extractive development. Even though each of the projects discussed here were at early stages of assessment and outcomes were far from being realised, the experiences explored in the two parts of the case study highlight that justice is a concern across the full spectrum of environmental decision making.

Essentially, conflict arose in the Namoi catchment because critical place-based concerns were either excluded or not responded to by those in power. Stakeholders interviewed felt that, overall, the nature of the opportunities for public input and review of extractive development had left them without 'a seat at the table' (Adam) and the lack of recourse to review decisions meant that the system was viewed as 'not fair and equitable to all people involved' (John). As Eric observed, decision makers failed to 'listen to the people who are getting impacted':

> [Y]ou know, they set up a process for the people to be heard, but there's no process to check if anything that's heard is listened to.

Efforts to challenge existing circumstances were counteracted by the influence of proponents and industry who were able to frame extractive development as critical to the state's economic prosperity and energy security. Campaigns such as the NSW Minerals Council's 'Hurt Mining, Hurt NSW'[10] achieved political prominence by invoking the benefits of extractive development to the broader public good. The evidence is that this industry rhetoric was subsequently interwoven with government discourse and used to justify governance interventions which further disempower those opposed to development. As Susan summarised of the Namoi catchment case:

> There's a key word missing in any of the governance at the moment . . . integrity.

Notes

1 Friends of the Pilliga 2011b. 'Clayton's' is an Australian colloquialism which arose from a popular advertising campaign in the 1970s and 1980s for a non-alcoholic beverage, 'Clayton's', sold in a container resembling a bottle of alcohol and marketed as 'the drink you're having when you're not having a drink'. The name has since come to mean 'something that is illusory or exists in name only; a poor substitute or imitation' (Robinson 2013).

2 Unlike mineral exploration activities, which do not constitute SSD under Part 4 of the *Act*, and are permissible without development consent: *State Environmental Planning Policy (Mining, Petroleum Production and Extractive Industries) 2007* (NSW), Reg 6; *State Environmental Planning Policy (State and Regional Development) 2011* (NSW), sch. 1, cl. 5.

3 *State Environmental Planning Policy (State and Regional Development) 2011* (NSW), Sch. 1, cl. 6 (2) (c) (as at 28 September 2011).

4 *State Environmental Planning Policy (State and Regional Development) 2011* (NSW), Sch. 1, cl. 6 (4) (as at 28 September 2011).

5 *Connell v Santos NSW Pty Ltd* [2014] NSWLEC 1.

6 *Environment Protection and Biodiversity Conservation (EPBC) Amendment (Standing) Bill 2015* (Cth).

7 Under the *State Environmental Planning Policy (Infrastructure) 2007* (NSW).

8 *People for the Plains Incorporated v Santos NSW (Eastern) Pty Limited and Ors* [2016] NSWLEC 93 (1 August 2016).

9 *Petroleum (Onshore) Act 1991* (NSW) Sch 1B.

10 See www.nswmining.com.au/people/hurt-mining-hurt-nsw

6   The Marcellus Shale
       case study

Introduction

The case study discussed in this chapter is set in the state of Pennsylvania in the United States. Pennsylvania is a state rich in natural resources, including the largest shale gas formation, the Marcellus Shale (see Figure 6.1), in the US. Estimates are that the Marcellus Shale can provide national energy security as well as economic growth for the state – particularly in the largely rural areas where gas drilling activity is concentrated. The state has a long history of extraction, but the rapid expansion of the industry over recent years has triggered many concerns, particularly at the local scale with respect to environmental and social impacts.

Unlike the previous case studies, which concentrated upon the fates of specific development projects, the focus of this case study is on the processes of governance formulation and implementation, and in particular, the ways in which the development of a new regulatory framework for oil and gas development reinforced government and industry power in an effort to centralise land use decision making.

As with the previous case studies, this chapter draws on media reports, policy documents, legislation (see Table 6.1), case law and interview data (see Table 6.2) to explore perceptions of justice and fairness in decision making processes relating to shale gas development.

Unconventional gas extraction in the Marcellus Shale Formation

Pennsylvania's history, as Brian (see Table 6.2) put it, is 'punctuated by waves of resource extraction'. Since the 1700s, the state has been a major producer of natural resources, including timber, coal and steel. Pennsylvania is also the birthplace of the American oil and gas industry, with the first commercially successful oil well drilled in the state in the mid-1850s (Pifer 2010). By the early twentieth century, Pennsylvania was producing one-half of the world's oil supply (Eisenberg 2015). However, the eventual expansion of exploration activities elsewhere ultimately reduced the industry in Pennsylvania to a 'faint pulse' by the turn of

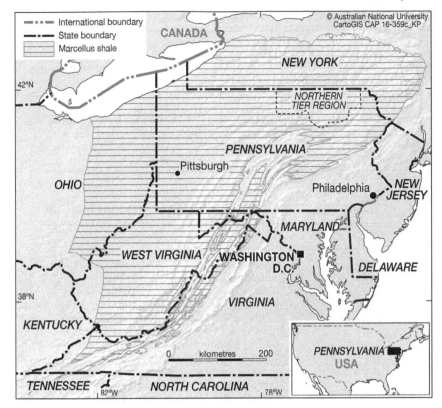

Figure 6.1 Location of the Marcellus Shale Formation

Table 6.1 Regulatory framework relevant to the Marcellus Shale case study

Law	Function
Oil and Gas Act (Act 223 of 1984, 58 Pa. Cons. Stat. §§601.101–601.607)	Governs the development of oil, gas and coal in Pennsylvania, including environmental standards, and conduct of drilling operations.
Oil and Gas Act (Act 13 of 2012, 58 Pa. Cons. Stat. §§2301–3504) ('*Act 13*')	Revoked and replaced the 1984 *Oil and Gas Act*.
Pennsylvania Municipalities Planning Code (Act 247 of 1968, P.L. 805; 53 P.S. §10101)	Provides counties and municipalities with the authority to enact local land use ordinances.
Federal Water Pollution Control Act (Clean Water Act) 1977 (33 U.S. Code 1251–1387) ('*Clean Water Act*')	Regulates the integrity of the nation's waters.
Comprehensive Environmental Response, Compensation, and Liability Act 1980 (42 U.S. Code §9601) ('*CERCLA*')	Prescribes specific processes for the investigation and clean-up of sites contaminated with hazardous wastes. The *CERCLA* is administered by the EPA.

(*Continued*)

Table 6.1 (Continued)

Law	Function
Safe Water Drinking Act 1974 (42 U.S. Code §300f)	Regulates the protection and quality of drinking water.
Resource Conservation and Recovery Act 1976 (42 U.S. Code ch. 82 §6901)	Governs the disposal of hazardous wastes.
Emergency Planning and Community Right to Know Act 1986 (42 U.S. Code §11001–11050)	Governs emergency response preparedness and reporting requirements for toxic chemical usage.

Table 6.2 Stakeholders interviewed for the Marcellus Shale case study

Name*	Role
James	County government representative/landholder in shale gas area
Sean	County government representative
Sandra	County government representative
Christopher	County planning department representative
Frank	County planning department representative
Gary	County planning department representative
Jeff	County conservation district representative
Martin	Federal–interstate environmental authority representative
Ryan	Environmental advocacy organisation representative
Margaret	Environmental advocacy organisation representative
Steven	Environmental advocacy organisation representative
Brian	Environmental advocacy organisation representative
Benjamin	Environmental advocacy organisation representative
Edward	Environmental advocacy organisation representative
Joseph	Lawyer, environmental advocacy organisation
Roger	Lawyer, environmental advocacy organisation
Matthew	Legal advisor
Linda	Community group organiser
Carl	Economic development organisation representative
Ronald	Extension officer

* All names listed here are pseudonyms

the century (Smith 2011: 4). By the early 2000s, the combination of increasing energy prices and the availability of new technologies prompted a revival of Pennsylvania's oil and gas industry (Pifer 2010: 49; see also Smith 2011; Fershee 2014). As Smith and Ferguson note, '[t]housands of gas workers flowed into the state from around the country, and an army of trucks, moving equipment, extracted gas and water descended upon the largely rural landscape' (2013: 379). Natural gas production has since quadrupled in Pennsylvania (Cosgrove, LaFave, Donihue and Dissanayake 2015).

BOX 6.1   The Marcellus Shale Formation

The Marcellus Shale Formation is a unit of sedimentary rock which lies beneath the Appalachian Mountain range across the states of Pennsylvania, New York, West Virginia, Ohio and Maryland in the US. Approximately 64 per cent of Pennsylvania lies over the Marcellus Shale (Clough and Bell 2016). Located 1,200 to 2,400 metres below the surface, the Marcellus Shale holds vast natural gas reserves, so much so that it is the largest shale gas formation in the US, and one of the largest in the world. Some have estimated that the Marcellus Shale holds sufficient gas reserves to meet the heating and energy needs of the US for as much as 45 years (Sovacool 2014), and will lead the US to become the world's largest exporter of natural gas by 2020 (Mondock 2015).

Shale gas exists within sedimentary rock and, due to its low permeability, typically requires unconventional drilling processes, such as horizontal drilling and hydraulic fracturing (fracking), to stimulate and release the gas. Unlike conventional drilling, which is generally able to access naturally flowing gas reserves through vertical wells, the process of horizontal drilling involves drilling both vertical and lateral well bores to access gas (Wiseman 2010; Jefferies 2012; Sovacool 2014). Hydraulic fracturing may then be used to expand the fractures in the shale surrounding the drilled well to enable the gas to flow from the well. This involves the injection of large volumes of fracturing fluids (a mixture of water and chemicals) and sand into the well at high pressure to act as a proppant to hold the shale fractures apart (Sovacool 2014). Natural gas then flows back up to the surface through the well bore, along with some of the fracturing fluid (flowback water), which is disposed of or reused for additional fracturing activities (Wiseman 2010).

Recovery of gas from the Marcellus Shale was previously not economically viable, but the advance of drilling and fracking technologies over the past 20 years, combined with higher gas prices, has made the Marcellus Shale not only cost-effective but highly profitable (Wiseman 2010; Smith 2011; Jefferies 2012; Anderson 2013; Sovacool 2014). By the end of 2015, it was estimated that over 9,500 wells had already been drilled in Pennsylvania (Clough and Bell 2016), with potentially up to 100,000 wells to be drilled across the multi-state Marcellus region (Schafft and Biddle 2015). Of the 6 million people residing in areas covered by the Marcellus Shale in Pennsylvania, approximately 3.5 million are located in rural counties, many of which have a population density of less than 100 people per square mile (Clough and Bell 2016).

Much of the gas extracted to date from the Marcellus Shale comes from the Northern Tier region of Pennsylvania, particularly from Tioga, Bradford, Susquehanna and Sullivan counties (Higgins 2013). The Northern Tier spans 4,000 square miles and is home to just under 180,000 people, who reside in small rural communities (Northern Tier Regional Planning and Development Commission n.d.). The region has historically experienced higher unemployment rates than other counties throughout Pennsylvania and per capita incomes have been among the lowest (Northern Tier Regional Planning and Development Commission n.d.). Timber and agriculture, particularly dairy and veal production, were traditionally dominant industries but have since declined. As these industries began to wane and unemployment levels rose throughout the region, the shale gas industry escalated, actively encouraged through regional development incentives (Eaton and Kinchy 2016).

Supporters of natural gas extraction argue that the industry can provide energy independence and security along with a range of economic benefits (Jefferies 2012; Wagstaff 2012; Apple 2014; Fershee 2014; Clough and Bell 2016; Sovacool 2014; Willow 2016). Proponents also argue that shale gas has a cleaner environmental footprint compared with other fossil fuels resulting in lower emissions than coal or oil fired power generation, potentially enabling it to operate as a 'bridge fuel' to transition to cleaner power sources (Argetsinger 2011; Sovacool 2014). A further argument advanced in favour of the shale gas industry is the promise of economic growth and jobs; a 'godsend' to rural communities suffering from high unemployment and stagnant economic conditions (Wagstaff 2012: 328; Wilber 2012). Sean summarised the views of a number of interviewees:

> Jobs . . . it's a huge and a very compelling argument especially in Pennsylvania, you know, steel mills are gone, farming has gone downhill, we have a lot of economic stress in this state and [the Marcellus Shale] is seen by most as the salvation.

However, the shale gas revolution has raised many concerns, especially regarding human health and environmental impacts, particularly with regard to risks to the quality and quantity of water resources (Jefferies 2012; Argetsinger 2011), along with air pollution, threats to biodiversity and potential contamination from flowback leakage and industrial accidents (Perkins 2012; Sovacool 2014). Various chemicals used in hydraulic fracturing include either known or suspected toxins, carcinogens or endocrine disrupters (Fisher 2015). Some research has linked negative human health impacts with close proximity to shale gas development, including neurological disorders, respiratory problems, frequent headaches and nosebleeds, pregnancy complications and miscarriages, as well as various cancers (Rahm 2011; Clough and Bell 2016; Sangaramoorthy et al. 2016). The negative social impacts of fragmented landscapes and disruptions caused by large-scale industrial drilling practices – which can consist of up to ten wells per well pad (Smith 2011; Kitze 2013) – include the capacity of rural communities and infrastructure to cope with increased industrial activity (such as noise and light pollution, and traffic) and the social effects of a transient workforce and potentially unequal access to economic benefits and housing (Brasier et al. 2011; Perkins 2012; Sangaramoorthy et al. 2016).

The promised economic benefits of shale gas development have also been disputed, particularly when offset by longer term social and environmental impacts (Andrews and McCarthy 2014). Clough and Bell (2016) found that income levels have remained relatively stable since the early 2000s for those living in close proximity to Marcellus Shale gas development. Wrenn, Kelsey and Jaenicke (2015) found that there have been only moderate increases in local resident employment, with a significant number of jobs going to people who reside either out of the county or the state. Hardy and Kelsey (2015) found that, although shale gas development increased resident lease and royalty income, the distribution of that revenue has been concentrated among a small proportion of residents. One county commissioner interviewed for this research, James, noted that there had been a boom in local business but that many skilled jobs were being filled by non-local workers; and that the influx of transient workers, who were renting hotel rooms on a long-term basis, had reduced the amount of hotel occupancy tax revenue (because stays longer than 30 days were exempt from the tax) while simultaneously stretching local resources, including emergency and social services, and even courts and correctional facilities. Clough and Bell conclude that there is 'no evidence that there have been widely diffused benefits for people living near well sites' (2016: 7) and that the shale gas boom in Pennsylvania has not delivered on its promises of broad economic growth. As Ryan, an environmental NGO representative interviewed for this research, remarked, the 'economic argument carries the seed of its own repudiation which is . . . the bust . . . its boom and bust'.

Subsurface mineral rights ownership issues further muddy this already difficult conundrum. Unlike Australia, where subsurface mineral and gas rights are retained by the government, the US allows private ownership of subsurface rights. A landholder may thus lease subsurface gas rights to a drilling company in return for financial compensation (Anderson 2013).[1] However, landholders may not necessarily be in possession of those subsurface rights. Mineral rights may be severed from surface rights as a 'split estate', and passed to another party who can then lease them. Severance of mineral estates is common in Pennsylvania and, in many instances, interests have been split by previous surface owners – sometimes generations ago (Anderson 2013; Pifer 2010). This has created a situation where not all surface estate owners possess rights to the oil and gas underneath their land and they may be limited in controlling or limiting drilling activities (Pifer 2010). Because mineral estates are dominant over surface estates, and mineral rights owners are entitled to 'reasonable use' of the surface to develop their mineral rights, surface owners may be left with little remedy against mineral owners who seek to develop oil or gas interests on their land (Anderson 2013; Andrews and McCarthy 2014). In some cases, surface owners have been unaware that they did not own subsurface mineral rights; as Sandra, a county government representative interviewed for this research, noted, 'it was a huge shock to a lot of people that thought their family had had these rights for the last 100 years and then found out that they were reserved'. In certain areas of the state, drilling companies had already purchased a substantial portion of the available mineral rights – in some cases 'years before fracking had even become a publicly discussed issue' (Apple 2014: 243).

Several interviewees for this research expressed concern about how mineral estate owners had been held to account for impacts upon the surface, particularly given that shale gas drilling operations were quite intensive and invasive. Sean, a county government representative interviewed, shared this story from his municipality:

> There's a case which took place not far from here . . . a property owner who owned the surface rights and not the subsurface rights had a very large intrusion upon his property with the felling of trees, the fragmentation of the habitat and he was helpless to have any control over that . . . The company was well within its rights to drill and harvest that gas and the company is required to restore the property, but his response to that is 'How are you going to restore an 80-year-old forest?' . . . that still troubles me that the law will create a situation like that.

While surface owners still in possession of mineral rights have been able to sign (in some cases) quite lucrative drilling leases and direct the conduct of drilling activities on their land (Andrews and McCarthy 2014; Pifer 2010), others have not necessarily been free to exercise such choice (Apple 2014; Sangaramoorthy et al. 2016). Describing the vulnerability of low-income municipalities to uncontrolled shale gas development as an 'environmental and economic injustice', Apple notes that the 'legal–economic system and the reasoning behind it often disguise inevitable outcomes of unequal bargaining as free choices within a free market when, in reality, it has created a system in which it is impossible to resist the pressures of economic need' (2014: 234). Frank, a county planning department representative interviewed for this research, agreed:

> What we found here is that while there are a few people who are philosophically against it and won't sign [a drilling lease], most people are saying, 'You know, I'm struggling to make ends meet'.

'Rampant leasing', as Brian put it, escalated across many communities in Pennsylvania (particularly in the Northern Tier), with 'a lot of high pressure sales going on'. Sean remarked that 'it was modern day gold rush', and the 'land men' who negotiated leases were considered by many interviewees to be 'pretty unethical' and 'one step lower than a used car salesman' (Jeff), through their use of aggressive tactics to secure leases from landholders (see also Wilber 2012). The land men were seen by interviewees to have done a 'good job of telling people what they want to hear', which had, in some cases, 'really created some hardships' (Christopher).

Public concern about shale gas development intensified, arguably fuelled by media reports of contamination incidents, as well as documentaries such as Josh Fox's *Gasland*, which garnered celebrity attention and brought fracking to a wider audience (Spence 2012; Fershee 2014; Kitze 2013).[2] Industry groups challenged the veracity of Fox's documentary, responding with the film-based rebuttal *Truthland*.[3] Interviewees for this research noted that the 'duelling documentaries' did little to unravel the complexities surrounding the development of the Marcellus Shale, and that the absence of 'good facts' impeded the ability to make sound policy decisions and resolve community concerns. As the gas industry continued to boom throughout Pennsylvania, and communities faced an increasingly problematic distribution of environmental and social outcomes, the regulation of Marcellus Shale development became a principal focus of public debate.

Regulation of the Marcellus Shale

Regulation of gas development in the Marcellus Shale crosses federal, state and local government levels. There are no comprehensive federal regulations concerning shale gas development, but several federal laws prescribe certain minimum standards for particular aspects of it (Kitze 2013; Spence 2012). In particular, the *Clean Water Act* prohibits the discharge of pollution into waterways, and the *Comprehensive Environmental Response, Compensation, and Liability Act* establishes liability for the contamination of well sites (Wiseman 2010; Jefferies 2012). Overall, though, the federal regime is 'most notable for what has been exempted or not regulated' (Jefferies 2012: 98) (see Box 6.2).

BOX 6.2 Exemptions for shale gas development under federal regulation

- The *Safe Water Drinking Act*, which was established to regulate drinking water across the nation, authorises the federal EPA to regulate for the protection of groundwater drinking supplies including the regulation of hydraulic fracturing as an 'underground injection' that could impact drinking water sources (see *Legal Environmental Assistance Foundation v EPA*, 276 F.3d 1253, 1256 (11th Cir. 2001). In 2005,

---

Congress amended the *Act* to expressly exclude hydraulic fracturing from the underground injection requirements (Wiseman 2010; Jefferies 2012; Spence 2012; Higgins 2013). Congress also provided for further exemptions for oil and gas activities from storm water runoff regulations (Andrews and McCarthy 2014).

- The *Resource Conservation and Recovery Act*, which governs the disposal of hazardous wastes, exempts hazardous wastes resulting from oil and gas exploration and production (Wiseman 2010).
- Oil and gas operators are exempted under the federal *Emergency Planning and Community Right to Know Act* from reporting on annual toxic chemical releases occurring during fracking operations (Wiseman 2010; Jefferies 2012; Spence 2012).
- Other exemptions also exist for shale gas extraction in national standards for emissions of air pollutants (Andrews and McCarthy 2014).

The limited federal oversight leaves states largely responsible for filling the void of regulation with respect to shale gas exploration, including such matters as the development of well sites, disposal of fracturing fluid, control of water impacts, data collection and reporting, and prevention of spills and leaks (Wiseman 2010; Jefferies 2012; Fershee 2014). Some Marcellus jurisdictions, such as New York, responded to the growing conflict over development by implementing outright bans on hydraulic fracturing (Fershee 2014). Pennsylvania, on the other hand, forged ahead with shale gas exploration under existing legislation – namely, the *Oil and Gas Act* of 1984, which was 'intended to promote the development of the state's oil and gas resources' (Foote 2015: 245) through the regulation of exploration, development and production. State laws are the principal mechanism for regulating the impacts of gas extraction in the US but many (including environmental laws) 'do not adequately address . . . community interests' that arise in the context of shale gas drilling (Stares, McElfish and Ubinger Jr 2016: 103). Indeed, Pennsylvania's existing legislative framework was ill-equipped to deal with the concerns emerging. As Browning and Kaplan have argued, it was 'weak': it did not levy any severance taxes (unlike other oil and gas producing states), it insufficiently funded enforcement operations, and left Pennsylvania as 'the only state to allow drillers to dump fracking wastewater in its waterways' (2011: 19). Interviewees for this research similarly observed that the *Oil and Gas Act* 'didn't even contemplate hydraulic fracturing, or the scale of activity' (Steven), which rendered efforts to regulate shale gas development within the existing legislative regime 'like trying to build a Corvette on a Model T line' (Roger).

Pennsylvania's legislature was seemingly 'caught by surprise by the pace and extent of the Marcellus development' (Andrews and McCarthy 2014: 11). Margaret and Roger believed that Pennsylvania's 'extractive mentality', forged by a long history of resource removal where extraction had become part of the state's 'culture and life blood', had generated inertia with respect to state-level regulatory action. As Steven summarised: 'Because we have such a history of [extractive] development the legislature is very hesitant to try to pass anything that would restrict that.' Brian concurred: 'Our history is resource extraction so we're used to getting raped. I mean, I grew up with orange and turquoise streams from acid mine pollution.'

Other interviewees suggested that a succession of pro-shale gas governors, namely Ed Rendell (2003–2011) and Tom Corbett (2011–2014), had actively supported a permissive regulatory regime which facilitated the rapid growth of the shale gas industry. Interviewees felt that Governor Corbett, in particular, had 'coarse ties to industry' (Roger), and that his administration had been 'very supportive' of shale gas development 'to the point of giving the appearance that they didn't potentially care about some of the impacts or the feelings that others might have' (Ronald). Non-profit advocacy group, Common Cause, have argued that government support for shale gas development was fortified by generous campaign donations from the oil and gas industry in appreciation of Pennsylvania's lenient regulatory stance (Browning and Kaplan 2011). They report that by the end of 2010, Governor Corbett had become the state's leading recipient of campaign contributions from the oil and gas industry, totalling US$1.6 million (Browning and Kaplan 2011: 19). Campaign contributions from the oil and gas industry to other candidates, parties and officials in Pennsylvania totalled US$6.1 million between 2001 and 2010, as the industry 'came to recognize the potential profitability of Pennsylvania's unique geographical and regulatory positioning' (Browning and Kaplan 2011: 21).

Several interviewees noted that intensive marketing campaigns had further reinforced a perception of the necessity and legitimacy of shale gas development in Pennsylvania. As Brian remarked, the industry 'spent billions of dollars on television commercials . . . it became the conventional wisdom that the industry created several hundred thousand jobs during the recession'. Sica has similarly observed that extensive advertising framed shale gas development to be of benefit across a number of scales, both within and beyond the state:

> Radio advertisements paid for by America's Natural Gas Alliance entreat listeners to 'think about' the transition from other fossil fuels to natural gas. Uncredited billboards along the highways broadcast messages like 'Sun Sets, Wind Stops – Natural Gas: Reliable, American Energy'. The ubiquitous TVs and electronic devices display 'energytomorrow.org' ads, which portray fracking as a public good . . . Shale gas appeared to provide income, savings or general aid to wider sections of the public than just oil and gas firms, the state and a subset of rural landowners who leased their land for fracking.
>
> (Sica 2015: 443–444)

Local governments[4] throughout Pennsylvania were left to play regulatory 'catch up', as Joseph (an environmental lawyer interviewed for this research) put it, in order to deal with the mounting conflicts over shale gas development within their boundaries. The primary method by which county and municipal governments can control oil and gas development is by dividing land according to zones and then permitting or restricting certain land uses (e.g. residential, industrial, agricultural) in each zone (Smith 2011; Smith and Ferguson 2013; Kitze 2013). However, zoning authority is delegated by the state, and there is only limited capacity for local governments to use zoning to restrict gas activities (Stares et al. 2016). Specifically, former §602[5] of the *Oil and Gas Act* pre-empted (see Box 6.3) local government efforts to regulate drilling operations otherwise covered by the *Act* (Foote 2015; Stares et al. 2016).

BOX 6.3   Pre-emption

Pre-emption is 'the simultaneous expansion in power of a higher level of government and reduction in power of a lower level of government' (Weiland 1999: 468), which may be used to establish uniform state-level regulation. As Wagstaff explains, pre-emption 'provides a hierarchy for potentially conflicting laws that governs when broad local powers . . . come into conflict with more specific state powers like the *Oil and Gas Act*' (2012: 339).

Former §602 of the *Oil and Gas Act* provided that local ordinances and enactments pursuant to the *Pennsylvania Municipal Planning Code* of 1968 could not 'impose conditions, requirements or limitations on the same features of oil and gas well operations regulated by this act or that accomplish the same purposes as set forth in the act'. Section 603 (b) of the *Municipal Planning Code* reiterates the pre-emption imposed by the *Oil and Gas Act*. Essentially, these pre-emption provisions meant that municipalities could 'dictate some of how drilling and associated development activities go forward, but not whether they go forward' (Andrews and McCarthy 2014: 12).

Nonetheless, these restrictions did not discourage some communities. In an apparent 'effort to protect citizens from the unseemly side of mineral development' (Foote 2015: 245), many municipal governments exercised local zoning authority in order to control or limit the operation of drilling activities within municipal boundaries (Perkins 2012; Wagstaff 2012).

Exerting local control

By 2011, over 800 counties and municipalities had adopted local zoning regulations and ordinances to restrict noise, road use and setbacks of drilling activities (Marcellus Shale Advisory Commission 2011). But, as several interviewees indicated, pre-emption limited what local governments could do: 'we did what we could locally, but we were in those narrow parameters [of pre-emption] so we couldn't do a lot' (Frank). Roger captured the view of many interviewees who

believed that the threat of pre-emption litigation led many municipal governments to draft quite conservative ordinances:

> Who is more likely to sue you if you make a decision in an oil and gas zoning
> activity? The oil and gas company . . . because they've got the deep pockets
> and they can fund litigation . . . I think that rural municipalities in general
> were more likely to side with gas industries even if they had reservations
> about what they were doing . . . because they're afraid of being sued.

However, other municipalities tested the boundaries of local zoning authority,
going as far as implementing local moratoria and 'symbolic resolutions in support
of anti-drilling legislation' (Andrews and McCarthy 2014: 12). Sandra remarked
that some municipalities had adopted 'unenforceable and unconstitutional ordinances prohibiting drilling'. Many local zoning ordinances thus became the subject of litigation with gas companies seeking to challenge their legitimacy, and
the courts quickly became congested with pre-emption cases.

Two Supreme Court cases decided in 2009 looked specifically at the validity
of local zoning ordinances and the interpretation of §602 of the *Oil and Gas Act*:
Range Resources Appalachia LLC v Salem Township,[6] and *Huntley & Huntley, Inc.
v Borough Council of the Borough of Oakmont*:[7]

- *Range Resources* concerned a comprehensive municipal regulatory scheme
  (*Salem Township's Subdivision and Land Development Ordinance*), which covered several aspects of surface development relating to drilling, including
  permitting procedures for gas wells, regulation of well heads, pre-testing of
  water sources within 1,000 feet of well sites, site restoration post-drilling
  operations and a requirement to restore nearby streets to pre-drilling conditions (irrespective of whether road damage was caused by drilling vehicles).
  The Court found that the ordinance was not only 'a regulatory apparatus
  parallel to . . . the [*Oil and Gas*] Act', but that it acted as an 'obstacle to the
  legislative purposes underlying the Act' (at 877). Because the *Oil and Gas
  Act* and its regulations already dealt with many of the matters covered by the
  ordinance, it was pre-empted and thus invalid.[8]
- *Huntley & Huntley*, on the other hand, considered whether a municipality
  could designate zones for the location of gas wells. The Borough of Oakmont's zoning ordinance restricted gas drilling within districts zoned as
  residential, permissible only as a conditional use. Huntley & Huntley, Inc.
  lodged an application to commence drilling on two parcels of land located
  in a residential area, and the borough council refused to grant a conditional
  use permit for drilling to proceed, maintaining that the *Oil and Gas Act* did
  not pre-empt their ability to restrict the location of drilling activities. The
  Supreme Court agreed with the council's position, finding that the pre-emption provisions of the *Oil and Gas Act* did not take away the 'core municipal
  function' of local government to determine where certain land uses could
  occur. The Court ruled that the ordinance's restriction of oil and gas wells in
  residential districts was thus not pre-empted by the *Oil and Gas Act*.

Effectively, the two decisions created 'an upper and lower boundary for determining whether a local ordinance is pre-empted by the *Oil and Gas Act*' (Wagstaff 2012). Specifically, they confirmed that local governments could not directly regulate the processes of gas drilling and production already covered under the *Oil and Gas Act* through zoning ordinances, but could 'limit the locations' of drilling activities (Fershee 2014: 827). Some municipalities sought to capitalise on the decision in *Huntley & Huntley*, and industry complaints intensified that residential zoning ordinances were being used as a 'back door way to ban drilling altogether' (Foote 2015: 245; Fershee 2014). As Roger observed:

> Industry was not happy with the fact that they had to deal with hundreds of local governments with different laws, different cultures, they wanted consistency across the state. They lobbied very hard to the state legislature and to then Governor Corbett to get consistency.

In 2011, then-Governor of Pennsylvania, Tom Corbett, established the Marcellus Shale Advisory Commission to make recommendations on ways to reform the state's oil and gas laws in order to promote gas development, protect the environment and mitigate local impacts (Marcellus Shale Advisory Commission 2011: 8). The Commission was to be made up of 30 representatives from across state and local government, industry, conservation and farming sectors. However, from the outset, many viewed the Commission as imbalanced, with industry representatives far outnumbering other interest groups. Specifically, the composition of the Commission drew sharp criticism in light of the campaign donations made by many of its members; 14 members of the Commission (as well as their spouses) had made donations of a combined total of over US$440,000 to Governor Corbett's election campaign (McNellis 2011), and others had links with organisations and individuals who had also donated to Corbett's campaign (Mauriello and Olson 2011). One industry representative on the Commission had previously been litigated against by the Pennsylvania Department of Environmental Protection (DEP) for environmental violations at mines he owned (Mauriello and Olson 2011).

Several interviewees for this research believed that the Commission was tilted in the favour of the oil and gas industry, and that its members had been 'cherry picked' (Sandra) from industry supporters. The industry was considered to be 'very savvy about putting money into local politics' (Joseph), and knew 'how to get the votes in the right place'; '[T]here's unfettered campaign contributions in this state, the industry puts in huge amounts and they just figured they could buy it' (Benjamin).

From the perspective of environmental justice, public consultation and input into the preparation of legally binding rules for environmental decision making is a desirable participatory function (see Chapter 2). However, in the case of the Marcellus Shale Advisory Commission, interviewees perceived that the development of reform had been driven by the interests of industry, with scant regard for views expressed by those who had concerns about shale gas development.

The Commission held public meetings and visited several counties before compiling its recommendations for reform (Marcellus Shale Advisory Commission 2011: 9–11). After some four months of deliberations, the Commission released its final report, with 96 recommendations for the modernisation of oil and gas regulations that it claimed would 'ensure fair and consistent municipal regulation which does not unreasonably impede the development of natural gas' (Marcellus Shale Advisory Commission 2011: 114). Among the recommendations were the imposition of a drilling 'impact fee' for gas companies, increased penalties for violations of the *Oil and Gas Act*, and greater public disclosure obligations around hazardous substances and wastewater treatment and disposal. However, the recommendations were immediately deemed insufficient. Of chief concern was the imposition of an impact fee (as opposed to a direct severance tax), and the failure to include air pollution recommendations (which had been a particular issue expressed by communities) (McNellis 2011).

Pennsylvania's General Assembly began drafting amendments to the *Oil and Gas Act*, taking into account the recommendations of the Commission as well as recommendations made by a previous State Review of Natural Gas Environmental Regulations (Wagstaff 2012). By the end of 2011, both chambers had passed bills to amend the *Oil and Gas Act* and, on 14 February 2012, Governor Corbett signed into law *Act 13*, which revoked and replaced the 1984 *Oil and Gas Act*.[9] *Act 13* was a broad amending act which incorporated some of the Marcellus Shale Advisory Commission's recommendations (including the imposition of an 'impact fee') but, controversially, it also introduced an express pre-emption of local zoning for oil and gas development.

Act 13: 'Fair and consistent' regulation or 'fracking democracy'?

The new *Oil and Gas Act* had many similarities to its predecessor but also included several significant changes – not least of which was the inclusion of specific provisions relating to unconventional drilling activities (Wagstaff 2012). Specifically, *Act 13* introduced:

- a set of uniform state-wide standards concerning well permit approval procedures (§3211);
- restrictions on well placement and setbacks (§3215);
- protections for floodplains (§3215 (f)) and water supplies (§3218);
- disclosure requirements for the chemicals used in fracturing fluids (§3222); and
- the ability for municipalities to offer their views on any local 'conditions or circumstances' that should be considered by the DEP in relation to drilling permit applications under consideration by the DEP within their boundaries (§3212.1).

Beyond these specific land use and environmental provisions, two other key changes were implemented:

1. An impact fee for every well drilled in the state, the purpose of which was to ensure that producers are 'solely responsible' for the impacts of unconventional gas development (§3501(2)). *Act 13* provided counties with the ability to impose the fee on operators within their boundaries according to a prescribed levy (§2302(a)). The *Act* directs how the impact fee funds are to be distributed, with a portion going to various state agencies, and counties and municipalities receiving 60 per cent of the remaining funds (§2314). These funds must be used for purposes associated with natural gas development, including the construction and repair of roadways, bridges, storm water and sewage systems, public safety, environmental programs and social programs (§2314(d) and (g)). However, where a local ordinance is found to violate either the *Act* or the *Municipal Planning Code*, a municipality is deemed ineligible to receive any impact fee funds it would otherwise be entitled to, and remains so until the offending ordinance is amended or repealed (§3308).

2. The pre-emption of local ordinances purporting to regulate any matter covered by the *Oil and Gas Act*:

    a. §3302 provides that the state government 'preempts and supersedes the regulation of oil and gas operations', and §3303 further clarifies that local environmental ordinances which attempt to regulate oil and gas operations are also pre-empted.[10]

    b. §3304 (b)(1)–(3) expressly stipulates that local ordinances must provide for the 'reasonable development' of oil and gas resources. 'Reasonable development' means that municipalities must allow well and pipeline assessment operations (including seismic operations), and may not impose restrictions on the construction, fencing, structural height, lighting and noise of oil and gas operations that are more stringent than conditions imposed on other industrial uses within the zoning district.

    c. Furthermore, §3304(b)(5) provides that local governments must allow oil and gas operations (other than 'activities at impoundment areas, compressor stations and processing plants') as a permitted use in 'all zoning districts'. This means that activities associated with drilling operations – including hydraulic fracturing – must be allowed in all zoning districts, including residential areas.

    d. Impoundment areas are permitted in all zoning districts but compressor stations must be authorised in agricultural and industrial zoning districts, and processing plants must be authorised in industrial zoning districts (§3304(b)(6)–(8)).

    e. Municipal governments may not impose conditions on the hours of operation of drilling, compressor stations and processing plants (§3304(b)(10)).

Effectively, *Act 13* resolved the ambiguity which had arisen in the wake of *Huntley & Huntley*. It sought to cover the field of gas development by 'detailing with remarkable precision the contours of allowable land use ordinances' (Perkins 2012: 75). The amendments dealt specifically with the various local zoning ordinance tactics used by municipal governments to exert control over gas development activities by essentially requiring municipal governments to allow gas drilling in *all* zoning districts (Andrews and McCarthy 2014; Foote 2015). Consequently, *Act 13* generated considerable controversy and was labelled as 'fracking democracy' and the 'nation's worst corporate giveaway' (Rosenfeld 2012). The legislation fell short of total state control of gas development (which Governor Corbett had previously expressed a desire for), but it was nonetheless widely condemned for placing municipalities in a 'virtual straightjacket' that restricted their capacity to manage shale gas development within their borders (Perkins 2012: 77; see also Smith and Ferguson 2013; Apple 2014).

'Gamed from the beginning': flawed processes

Several interviewees for this research condemned the manner in which the legislation was developed:

> [*Act 13*] emerged from a conference committee at 5 pm on a Friday . . . it was written by a gas industry lobbyist, handed to the Republican Senatorial Leadership and they just put it out late one Friday afternoon to minimise the press . . . it was written by – and for – the gas industry.
>
> (Brian)

> People didn't know it was happening. It really happened overnight . . . There was the Republican House of Representatives and they sat down behind closed doors, you know, the industry basically drafted the law with the assistance of private practice attorneys who represented the industry, and other persons were really blocked out of the process.
>
> (Joseph)

Steven admitted that his environmental advocacy organisation had been consulted on the drafting of *Act 13*, but only on the portions that the legislature deemed of relevance to them:

> [T]hey developed pods of interest groups, so to speak. So we had the local government association, and the county association, they worked with them on the pre-emption issue, and the local control issue. They had a couple of us working on the environmental protection provisions, and another working group dealing with the impact fee. So we only really had some say on the environmental protection provisions . . . although we definitely had our opinions about it, we weren't asked about the impact fee or the pre-emption issue.

Roger detailed further:

> [I]t was gamed from the beginning, in so far as the administration picked its stakeholders and picked people from the environmental community who

would accommodate it more and object less to what it wanted to do. But even after, I mean the [Marcellus Shale Advisory] Commission ended up generating this set of recommendations [that] were very industry friendly. Some [recommendations] were pretty smart and would have been really good had they been implemented, but once they came out the Governor and the legislature ignored many of them. I say that's a good example of kind of gaming the system by starting up this public process that makes it seem very democratic . . . and then you get everybody's point of view and you listen to what you want to hear and ignore the rest.

The consultation process over the development of Act 13 highlights issues of procedural fairness. Interviewees perceived the decision making process to have been orchestrated in a way that limited broad public input, and provided greater opportunities for pro-industry voices to be heard. Even where concerns were able to be expressed, these were seen to have been largely ignored in the final draft bill. As Martin explained, people felt 'like no one's paying attention to them, no one's listening to them'. As we saw in Chapter 4, a disjuncture between the rhetoric and the realities of consultation, where feedback is seen to be overlooked or ignored, can lead to a perception of poor treatment and a sense of injustice. In the case of Act 13, the 'frustrating lack of opportunity for the Average Joe to influence the regulatory realm' was seen to have generated 'a lot of angst' and 'mistrust' of the government (Christopher). Overall, deficiencies in procedure and recognition were considered to have constrained the capabilities of citizens to shape the decision making institutions that were relevant to their very functioning and well-being.

'Manhandling municipalities': flawed outcomes

Beyond the way in which Act 13 was formulated, interviewees considered many of its outcomes to be unjust as well. In particular, the pre-emption provisions in §3304 were a major source of criticism, which were variously described as imposing a 'one size fits all state zoning requirement' (Roger), providing 'special treatment' (Frank) to the oil and gas industry, 'hijacked by economic interests' (Sean), and a 'terrible disservice to the citizens' (Sandra):

> Act 13 was passed in an effort to centralise all the decision making when it comes to gas exploration and extraction . . . it drew basically all the power and decision making process into state government and stripped away the authority of the local jurisdictions.
>
> (Edward)

> What [Act 13] basically did is it said 'local governments don't even get to control where anything is operated anymore, we're going to pre-empt them on that' . . . So you invest, I don't know, some decent sum of money in buying

a house and then the value of the house gets compromised by close-by shale gas drilling.

(Matthew)

The linkage of impact fee funding to compliance with the Act, including the pre-emption provisions, was considered distributively unfair. Interviewees condemned the broad authority granted to the Public Utility Commission (PUC) to assess compliance with the pre-emption provisions and questioned the adequacy of the hearing process that would take place to determine potential ordinance violations.[11] Interviewees also criticised the impact fee itself:

[The impact fee] was actually of benefit to the industry because they got out of it for a lot less than they would have had there been a severance tax.

(James)

Pennsylvania [has never had] a severance tax . . . because the industry started here and they had such control over the legislator that there was never a tax imposed.

(Christopher)

Specifically, the manner in which impact fee revenue was to be allocated led several interviewees to believe that it would not generate sufficient revenue to offset local impacts and legacy expenses:

What wound up happening is that the clamouring and the politics said no it should be delivered equally to everyone, so, predominantly, it was delivered equally to everyone with just a tiny part of it going towards those impact considerations.

(Jeff)

Joseph remarked that Act 13 was 'obnoxious' because it 'tried to manhandle local municipalities':

It actually commanded, it had express command in it that said, you know, you have to do this, you have to allow these facilities everywhere. But then there was a separate section that also said . . . if your ordinances don't comply with our command that you allow gas wells everywhere, you're also not going to get the impact fee.

Essentially, Act 13 positioned the state as the appropriate scale of governance for shale gas development in Pennsylvania. It reinforced the broad framing already undertaken by the industry in favour of shale gas development in Pennsylvania (as discussed earlier), while simultaneously disregarding the interests of those who were concerned with the costs of development at the community and even individual scale (Andrews and McCarthy 2014; Sica 2015). Interviewees felt that the dominance of the state scale in *Act 13* failed to recognise unique local concerns, and compromised the ability for local government to sufficiently control impacts at the 'nucleus of fracking development' (Apple 2014: 242). As Joseph noted:

> [Several townships] didn't like the fact that the statute was telling them what to do when they had spent time developing a comprehensive plan and to try to put zoning in place that would allow development in some rational way.

In terms of procedural fairness, many interviewees believed that *Act 13* further eroded already limited participatory opportunities for citizens. For example, in order to obtain a permit to drill, prospective drillers must apply to the DEP as well as send a copy of the permit application to surface landholders, relevant local governments and people with water supplies within 3,000 feet of the proposed unconventional drilling location.[12] However, municipalities have only 15 days to render their comments,[13] and the DEP is not obliged to consider the comments of municipalities in reaching its decision.[14] Surface landholders have only 15 days to object to an application and then only on the grounds that the application is incomplete or otherwise conflicts with the well location requirements set out in §3215(d).[15] If no objection is received, and the DEP does not object, then a permit must be issued within 45 days.[16] Several interviewees felt that both individual surface landholders and smaller municipalities would struggle to review and comment on permit applications within the allowed time period: '[communities] don't have that capacity' (Christopher); 'they really don't have the capacity to evaluate these plans and they certainly don't have the capacity to do it in a 15-day period' (Edward).

As emphasised a number of times in this book, the existence of opportunities to participate in environmental decision making does not necessarily translate into the capability to do so effectively – information provided to the public must be clear and unambiguous, and the public must have sufficient time to consider and comment upon proposals. Without these conditions, the capabilities of citizens to participate in environmental decision making may be constrained. Further, the lack of formal opportunities for participation in permitting decisions for those residents outside the 3,000 feet water supply boundary and the 25 per cent who rent property from a surface landholder (Kelsey, Metcalf and Salcedo 2012) raises additional issues of recognition. In such cases, local citizens often 'have relatively little voice in . . . decisions which have significant implications for their communities and for their own quality of life' (Kelsey et al. 2012: 13–14).

Several interviewees also believed that some key safeguards contained in *Act 13*'s provisions relating to environmental protection were illusory. For example, in considering a permit application, the DEP is bound by well location restrictions provided under §3215, which provides minimum setback requirements from water wells. However, under §3215(b)(4), the DEP is required to waive these requirements where a drilling operator submits a plan which identifies additional measures or practices that will be employed to protect water. Interviewees were particularly critical of this provision, and considered that it would enable the DEP to ignore the interests and concerns of other stakeholders. Brian summarised:

> The regulatory reforms that were embodied in *Act 13* . . . looked decent, but when you read the fine print, for example on setbacks, the DEP, the regulator, shall waive these requirements if the driller or owner of the mineral rights submits a plan. Not 'may' waive them, 'shall' waive them. So it's a paper tyre, a total paper tyre . . . The *Oil and Gas Act* has never been that strong and the rewrites of *Act 13* are what I would call 'eye wash' – they look decent on paper . . . but that language about 'shall grant waivers' essentially eviscerates it, so it's no good.

Some interviewees also expressed concern about the capacity of the DEP to adequately assess and enforce permit applications, citing reports which indicated that resource constraints had led the DEP to spend as little as 35 minutes considering development applications (Rubinkam 2011). Roger joked that the DEP now stood for the 'Department of Everything Permitted'. He went on further: 'It's a complete charade and there's no legitimacy at all. The DEP is not there to protect the environment, it's there to work with industry.' Edward similarly complained:

> Our Department of Environmental Protection, which has a lot of delegated authority from the National Environmental Protection Agency, has had reduced funding for the last decade . . . the resources that they have are not adequate to enforce the regulations . . . The last four years under the previous administration, they, frankly, were not allowed to do much enforcement.

These concerns were heightened by a subsequent executive order issued by Governor Corbett in the months following the introduction of *Act 13*, requiring the DEP to expedite the processing of permit applications (Corbett 2012). As Roger remarked:

> The permitting process that Pennsylvania has developed for oil and gas activities is a kind of junior varsity participatory process . . . It allows for virtually no public participation . . . [Oil and gas] permits are now issued on an expedited basis and no public notice is given of the application. You can find out through the DEP's website that it's been applied for but because they're processed on an expedited basis, practically there is no time to get in and look at the application before it's issued in most cases.

The combination of reduced funding for assessment and enforcement activities, the mandate to expedite the processing of applications and the limited framework for public participation each had a direct impact upon the capabilities of citizens to participate in the environmental decision making process, and control the outcomes of environmental decisions (Griewald and Rauschmayer 2014). Overall, many people perceived that Act 13 not only prevented municipalities from banning gas development but also left them 'with fewer bargaining tools to fight for responsible and safe development' (Apple 2014: 243). In the words of the interviewees:

> We got thrown a bone with Act 13 . . . some adjustments to the *Oil and Gas Act* established setbacks from streams and residences and roads and infra-structure that were better than what we had, but it went the other way with the zoning limitations.
>
> (Christopher)

> Act 13 was quite a mixed bag. It had some good things in it . . . but then on the flipside, it also had a couple of things that we viewed as an overreach, particularly in relation to local land use controls. It also neglected a whole range of issues, like air [quality]. The way the law was written, it was pretty blind to specific local needs or conditions. So it set these arbitrary param-eters, in terms of what setbacks should be from certain types of equipment or operations, but didn't take into account things like placement of schools, placement of hospitals, residential [areas] . . . it was a pretty broad swipe at trying to pre-empt everything.
>
> (Steven)

> The opportunity for the public to be involved [in land use decision making], now that the municipalities no longer have their zoning capabilities, is close to zero.
>
> (Benjamin)

Other commentators have since agreed that Act 13 was an inadequate attempt to regulate a difficult policy space: Foote considers that Act 13 failed to recon-cile the interests of stakeholders (2015: 257); Perkins concludes that both its processes and outcomes were 'sadly deficient in terms of their attention to sus-tainable development' (2012: 74); and Fershee is even more direct in his criti-cism, noting that Act 13 represented 'a heavy-handed, state-level response to local government actions to restrict or ban hydraulic fracturing that was not the ideal option to facilitate a proper balance between economic development and environmental protection' (2014: 861). Given the broad opposition to Act 13, many predicted that there would be swift litigation to dispute the legislation. Sure enough, within two months of its enactment, a constitutional challenge commenced in the Commonwealth Court of Pennsylvania.

The *Robinson Township* case: reclaiming local control

A coalition of seven municipalities, a doctor, two other individuals and the Dela-ware Riverkeeper Network, filed suit in the Commonwealth Court in March 2012 arguing that *Act 13* was unconstitutional. In July 2012, the en banc court (entire bench) voted four to three to overturn certain provisions of *Act 13*.[17] In particular, §3304 (concerning the pre-emption of local zoning ordinances) and §3215(b)(4) (concerning the DEP's ability to grant waivers from setback requirements) were declared unconstitutional.

The Court deemed that §3304 violated substantive due process under property rights provisions of the *Pennsylvania Constitution*. The majority judgement explained that §3304 essentially forced municipalities to permit gas operations in all zoning districts, requiring them to allow potentially incompatible land uses:

> §3304 violates substantive due process because it does not protect the interests of neighboring property owners from harm, alters the character of neighborhoods and makes irrational classifications – irrational because it requires municipalities to allow all zones, drilling operations and impoundments, gas compressor stations, storage and use of explosives in all zoning districts, and applies industrial criteria to restrictions on height of structures, screening and fencing, lighting and noise.[18]

The majority found that, in order to prevent the problem of incompatible land uses (the proverbial 'pig in the parlor instead of the barnyard'), zoning ordinances must satisfy due process. This requires them to be 'based on a process of planning with public input and hearings that implement a rational plan of development'.[19] The dissenting judges disagreed and believed that the majority had erroneously reached a 'legal conclusion that any zoning ordinance that allows a particular use in a district that is incompatible with the other uses in that same district is unconstitutional'.[20] They pointed out that the 'particular pig' of shale gas operations 'can only operate in the parts of this Commonwealth where its slop can be found' and that natural resources 'exist where they are, without regard to any municipality's comprehensive plan'.[21] They noted that the General Assembly, when drafting *Act 13*, had decided that it was in the best interests of citizens to ensure the development of gas resources. The majority, however, dismissed this, noting that regardless of whether oil and gas operations are classified as a 'pig in the parlor' or a 'rose bush in a wheat field', §3304 would create an 'unconstitutional "spot use"'[22] of zoning against the broader public interest.

The court also deemed that §3215(b)(4) violated the non-delegation doctrine under the *Pennsylvania Constitution*. The petitioners had contended that §3215(b)(4) had provided insufficient guidance to the DEP to waive the setback requirements, which did not satisfy the basic principles of the *Constitution* that legislation 'must contain adequate standards that will guide and restrain the exercise of the delegated administrative functions'.[23] Because the waiver authorised by §3215(b)(4) did not provide the DEP with any guidance as to how it might exercise its discretion or evaluate the plan submitted by operators, it was found unconstitutional as it gave the DEP the 'power to make legislative policy judgments otherwise reserved for the General Assembly'.[24]

The petitioners also claimed that other aspects of *Act 13* were unconstitutional and vague, but these arguments were all rejected by the Court and the relevant sections upheld. In particular, the petitioners had argued that *Act 13* had violated Article 1, §27 of the *Pennsylvania Constitution*. Otherwise known as the 'Environmental Rights Amendment' to the state's *Declaration of Rights* (analogous to the *Bill of Rights* in the US *Constitution*; Daly and May 2015), this section provides that people have a 'right to clean air, pure water, and to the preservation of the natural, scenic, historic and esthetic values of the environment' and, further, that 'Pennsylvania's public natural resources are the common property of all the people, including generations yet to come', and that the government must 'conserve and maintain them for the benefit of all the people'. Implemented in response to the negative history of extractive developments in the state (Dernbach, May and Kristl 2015) the provision was adopted by referendum in 1971, but rarely used in the intervening decades as a result of subsequent case law which imposed a specific balancing test for its application.[25] The petitioners argued that Chapter 33 of *Act 13* removed the ability for municipalities to balance oil and gas development and the preservation of the environment as required by the *Constitution*. The Court held, however, that Chapter 33 did not violate the *Environmental Rights Amendment*, because it effectively discharged municipalities of their constitutional obligation to 'strike a balance between oil and gas development and environmental concerns'.[26]

The day after the Court reached its decision, the government filed an appeal with the Supreme Court. However, the municipalities' success in the Commonwealth Court decision had 'emboldened other discontented localities and environmental groups' (Foote 2015: 258) who also filed *amicus curiae* briefs with the Supreme Court opposing the legislation. Interviewees for this research observed that the Commonwealth Court decision had created a greater awareness of the potential implications of *Act 13*, and had encouraged a groundswell of public support for the legal challenge against it:

> There's some backlash [against the Government] now that people have realised what the General Assembly have done. They now realise thanks to the [Commonwealth] Court and the people involved in that case.
>
> (Joseph)

> I think that *Robinson* was probably the beginning of the realisation that even small voices can be heard . . . like a David and Goliath-type situation where this little township was like, no, we're not going to stand for this, and stood up for what they thought was their rights and won.
>
> (Edward)

The Supreme Court heard arguments in October 2012 and handed down its decision over a year later, in December 2013. By a 4–2 majority, the Supreme Court affirmed the Commonwealth Court's determination that §3304 and §3215(b)(4) were unconstitutional, and found that §3303 was invalid.[27] While a majority of the judges agreed that the various provisions were unconstitutional, their reasons diverged. A plurality opinion was rendered by Chief Justice Castille, Justice Todd and Justice McCaffery,[28] who found that §3215(b)(4), §3303 and §3304 of *Act 13* were incompatible with the *Environmental Rights Amendment*, and thus invalid.[29]

Concurring that the respective sections were unconstitutional, Justice Baer opted to follow the Commonwealth Court's previous reasoning, and found that the breadth of *Act 13* violated substantive due process.[30]

Unlike the Commonwealth Court, the Supreme Court plurality found that the *Environmental Rights Amendment* contained in Article 1, §27 of the *Pennsylvania Constitution* did impose a fiduciary obligation upon the government to conserve and maintain natural resources. This obligation includes refraining from 'permitting or encouraging the degradation, diminution, or depletion of public natural resources', while also acting 'affirmatively to protect the environment, via legislative action'.[31] The plurality's particular interpretation of §27 was influenced by the legacy of environmental harm left by extractive industries in Pennsylvania.[32] Importantly, while noting that the economic well-being of citizens was 'obviously a legitimate state interest', the plurality opined that 'economic development cannot take place at the expense of an unreasonable degradation of the environment',[33] and that the *Environmental Rights Amendment* has the 'express purpose . . . to be a bulwark against actual or likely degradation'.[34]

Turning to the specific provisions of *Act 13*, the plurality found that while local governments are 'creatures of the state', §3303's pre-emption of all local ordinances relating to oil and gas 'went too far' (Pacheco 2015: 373). The 'blunt approach' of requiring industrial uses to be permitted in all zoning districts in §3304 was found to be 'incompatible with the Commonwealth's duty as trustee of Pennsylvania's public natural resources'.[35] The plurality considered that it would disable local government from 'mitigating the impact of oil and gas development at a local level', which may result in heavier burdens for certain properties and communities.[36] Because gas development is 'a quintessential local issue that must be tailored to local conditions', the plurality held that the impact of gas development on natural resources 'cannot reasonably be assessed on the basis of a statewide average'.[37] Finally, the requirement for the DEP to waive setback conditions under §3215(b)(4) was considered lacking in 'identifiable and readily-enforceable environmental standards', and would not ensure compliance with the requirement under the *Environmental Rights Amendment* to conserve and maintain the waters of the Commonwealth.[38] The plurality also noted that §3215 more generally 'marginalizes participation by residents, business owners, and their elected representatives with environmental and habitability concerns', encouraging decisions that are 'non-responsive to local concerns'.[39]

The plurality acknowledged that the constitutional challenge presented by the case was unprecedented;[40] indeed, their decision represented the first time that Article 1, §27 had been relied upon to declare a statute unconstitutional (Dernbach 2015). Nonetheless, the plurality justified overruling the various provisions of *Act 13*, because it would 'permit development with such an immediate, disruptive effect upon how Pennsylvanians live their lives'.[41] The plurality stated that:

> By any responsible account, the exploitation of the Marcellus Shale Formation will produce a detrimental effect on the environment, on the people, their children, and future generations, and potentially on the public purse, perhaps rivaling the environmental effects of coal extraction.[42]

In dissent, Justice Saylor took issue with the plurality's apparent findings of fact with regard to the detrimental impacts of gas development, which they relied upon to justify overturning the relevant portions of *Act 13*. He noted that the legislature 'possesses superior resources for information-gathering, debate, and deliberation in the policymaking arena', and that the court should not substitute its 'own policy preferences for those of the political branch'.[43] In particular, Justice Saylor criticised the plurality's 'non-record-based portrayal of *Act 13*'s impact', arguing that it had hypothesised 'an unreasonably deleterious impact of *Act 13* on the environment'[44] which was 'without a shred of evidentiary support'.[45] Further, by failing to recognise that municipalities are 'creatures of the General Assembly', created by (and deriving their power from) the state, Justice Saylor believed that the plurality had 'completely' redefined the role of municipalities relative to the state.[46] Justice Eakin, also in dissent, reiterated that municipal power to manage land use is derived from the state and may be removed or modified by the legislature without offending the *Constitution*.[47] The government sought review of the decision, which the Supreme Court subsequently denied.

Additional to their findings, the Supreme Court also remanded several unresolved issues to the Commonwealth Court for determination, focusing on the constitutionality of some of the remaining provisions from *Act 13* and whether they could stand independent of the invalid provisions.[48] The Commonwealth Court rendered its decision in July 2014, determining that the following sections remained valid:

- §3218.1, which limits the DEP's notice requirements in the event of chemical spills to public water facilities only;
- §3222.1(b)(10) and (11), which allows physicians to obtain confidential information about fracturing chemicals for the purposes of treating patients, but restricts their ability to disclose the information in order to protect its proprietary character (the so-called 'doctor gag rule'); and
- §3241, which confers the power of eminent domain on corporations to use private property to remove, store and transport natural gas for public use.

The Commonwealth Court was also under mandate to consider, in light of the Supreme Court's findings that §§3215(b)(4), 3303 and 3304 were unconstitutional, whether §§3302 and 3305–3309 were unenforceable. The Court found that §§3305–3309, which allow the PUC to review local ordinances to determine whether they violate *Act 13*, were now invalid because they were dependent upon the provisions that the Supreme Court had found unconstitutional.[49] However, the Court determined to uphold §3302, which provides that local ordinances must not regulate matters already covered by the *Oil and Gas Act* (such as those covered in Chapter 32).[50] In effect, this reaffirmed the previous decision in *Range Resources*,[51] that municipalities could not regulate the processes and operation of gas drilling and development, but could regulate where it takes place.

Through its reliance on the *Environmental Rights Amendment* of the *Pennsylvania Constitution*, the Supreme Court plurality's broad and novel opinion in

Robinson Township was one of the first decisions to raise environmental justice as a concern in the regulation of shale gas development (Pacheco 2015). As Daly and May note, it is 'certainly one of the few American cases from any jurisdiction that recognizes the comparability of social and collective rights with individual political rights' (2015: 163). Academic commentators have since offered comprehensive analyses on the potential implications of the decision, particularly its capacity to engage environmental constitutional rights in the US and globally (see Dernbach 2015; Daly and May 2015; Dernbach et al. 2015). However, while the Supreme Court decision in *Robinson Township* was undoubtedly historic and had 'significant persuasive power' (Dernbach 2015: 479), the fact that it was a plurality opinion (which does not form a binding precedent on other courts) initially diminished its impact.[52]

Nonetheless, litigation has continued with respect to other aspects of the decision. Gas pipeline construction emerged as the new battlefront over the Marcellus Shale, following the Commonwealth Court's finding that Act 13's conferral of eminent domain power on corporations to transport, sell or store natural gas under §3241 was valid.[53] With the infrastructure to deliver gas still in its infancy, pipeline project proposals have increased rapidly throughout the state over the last few years but have faced significant opposition. In order to construct pipelines, companies must negotiate with landholders to gain access to land and provide compensation for any damage or loss. However, in cases where a landholder refuses to negotiate, pipeline companies may be able to commence eminent domain proceedings to 'condemn' property, which enables them to acquire and use private property to transport, sell or store natural gas. Corporations may exercise the power of eminent domain where they satisfy public utility status, and possess a certificate of public convenience issued by the PUC.[54] Accordingly, several challenges against the public utility status of companies arose, questioning their capacity to assert eminent domain (and thereby avoid right-of-way negotiations with landholders).[55]

More recently, the Supreme Court heard appeals challenging the Commonwealth Court's previous determination on matters arising from *Robinson Township*, including eminent domain, the 'doctor gag rule', and the finding that the PUC could not review local drilling ordinances. In September 2016, the Supreme Court ruled:[56]

- that §§3305–3309 were not severable from the sections previously held to be invalid, agreeing with the Commonwealth Court's finding that the PUC could not review local ordinances nor withhold impact fee payments;
- that §§3222.1(b)(10) and (11) – the 'doctor gag rule' – was invalid, because it provided special treatment to the gas industry;
- similarly, that §3218.1, which required the DEP to provide notice only to public drinking water facilities potentially impacted by a spill, also be struck down as a 'special law' benefiting specific groups; and
- that §3241, which granted private companies the power of eminent domain to seize property for gas storage and transport, violated the US and Pennsylvania constitutions.

Interestingly, the Supreme Court declined the opportunity to revisit whether parts of *Act 13* violated the *Environmental Rights Amendment* of the *Pennsylvania Constitution*, despite the fact that this was central to the previous plurality opinion which struck down §3215(b)(4), §3303 and §3304. It thus remains to be seen how this matter will be decided in the future, and whether the *Environmental Rights Amendment* gains further momentum in resolving conflicts over shale gas development. As Matthew remarked: '[T]he court is not going to be able to just put the genie back in the bottle. It's in the *Constitution*, you can't undo that.'

In the meantime, other parts of the *Act* remain in force, including the impact fee, and the regulations governing drilling activities. Thus, local governments are now only limited by the *Municipal Planning Code* when seeking to utilise local zoning ordinances to control gas development.

Beyond *Robinson Township*: local control versus local capacity

In the wake of *Act 13*, it was the 'potential loss of power "to determine . . . the long-term character of their local communities" and to protect themselves from the risks and impacts of fracking development' that municipal officials most deplored (Apple 2014: 242).[57] Most certainly, it is local communities who 'experience the effects of fracking most intensely and profoundly' (Spence 2014: 412): the social and environmental costs of the increased industrial activity associated with shale gas development tend to be focused at the local level, while many of the benefits are spread beyond municipal borders. The *Robinson Township* litigation has undoubtedly returned control to local communities but, in doing so, has reignited the debate over who is the appropriate polity to regulate shale gas development.

Spence notes that local control can 'allocate the decision to those who care the most and who experience most of the impacts' (2014: 389); as well as act as an important bargaining tool, incentivising those who will benefit from shale gas production to fairly negotiate with those who have the power to withhold development consent to better share the gains and adequately compensate any losses. In reflecting upon their experiences with shale gas governance in Pennsylvania, interviewees for this research were generally supportive of some degree of local governance of shale gas development. Zoning was considered to be a traditional local prerogative that enabled municipal governments to control the impacts of development according to local needs and desires. Interviewees mentioned the importance of being able to respond to the distinct ways in which impacts were felt in rural communities, which were perceived to be frequently misunderstood by metropolitan decision makers:

> [T]hey don't understand the things that we have to face here . . . [for example] our roads are our lifeline, we have to use them to get to work, to get to

the hospital, to get to the food store. In Philadelphia, they can hop on a bus, hop on a train, ride a bicycle, we don't have that option . . . our roads are as important or more important but they don't take that into consideration because we're rural . . . it's a constant daily battle to bring to the attention of the elected officials, the unique challenges and issues that we face in rural Pennsylvania.

(Sandra)

While interviewees were generally pleased that *Robinson Township* had restored some local control over the impacts of shale gas development, some noted that municipal governments still did not wield much power. Specifically, interviewees observed that the decision had merely reinstated the position post-*Range Resources*, where municipalities had 'minimal power over *where* fracking can take place' (Kitze 2013: 402) but could not control *how* it took place. Several interviewees believed that local control should be strengthened beyond siting power in order to adequately manage the 'consequences of [shale gas development] at that very local scale' (Benjamin). For example, Gary remarked:

I believe you should be able to form your own regulations to guide your municipality the way you want it to develop. You don't want the state coming in and telling you how your municipality should develop.

However, even though local governments were considered to be 'important laboratories of regulation' for shale gas development (Wiseman 2014: 35), many interviewees considered that certain smaller municipalities lacked the capacity to engage effectively in local decision making:

[*Robinson Township*] really put zoning back in the locals' hands, but what it also did was give them a huge issue that they don't have a whole lot of capacity to deal with.

(Edward)

Each one of these little townships or municipalities, especially if they're in rural areas, they just don't have the money and they don't have the staff.

(Benjamin)

[I]t comes down to the capacity and the sophistication of the rural communities and [many] . . . don't have that training and education and ability.

(Christopher)

As discussed earlier in this chapter, even where municipalities have the ability to control fracking development through zoning, their overall power is dependent upon a range of factors, including bargaining skills and financial resources, and an understanding of the issues at stake (see also Apple 2014: 233). Several interviewees noted that large drilling companies were in a stronger position and could pressure municipalities to regulate in favour of gas development:

> You have a small township of a thousand people, they get an international corporation coming in with a troop of attorneys asking for a zoning variance or they'll sue you if they don't get it. And what's a township to do but give them a zoning variance?
>
> (Edward)

> [M]ost of these local level governments don't have much knowledge or expertise, or much planning capability. They're easily pushed around by the gas companies who will threaten to sue them for this, that or the other thing.
>
> (Benjamin)

This lack of power may be compounded by other obstacles specific to rural areas, including the scarcity of attorneys with sufficient knowledge of oil and gas law (Eisenberg 2015). The public interest environmental advocacy services that may be able to help in these circumstances 'get way more requests for assistance than [they] can possibly represent' (Edward). Other interviewees felt that even when communities were engaged with shale development issues, there was a lack of knowledge about how to defend local zoning against complicated pre-emption claims:

> [T]he number of people who actually have read up on pre-emption and how it works . . . and all the other things that go into pre-emption is pretty small, because it's a lot of archaic legal crap that people don't want to read.
>
> (Linda)

As discussed in the previous case study chapters, to ensure environmental justice it is necessary to address issues of inadequate access to information and resources (including legal advice), limited education and physical distance from centres of governance.

Enhancing capabilities: from 'winners and losers' to 'participants in a partnership'

Interviewees believed that community capabilities to participate in shale gas development decision making could be improved through more comprehensive and meaningful collaborative governance arrangements in cooperation with state and federal governments. Specifically, some interviewees felt that local regulatory efforts could be successful if they were underpinned by strong state and federal regimes. As Brian explained:

> There's a big controversy in this country: who should regulate shale and gas – the federal government or state government? And my response to that is

both. The federal government needs to set a floor beneath which nobody falls and then the individual states based on their geology and ecology need to add additional protections, not lessen . . . Then, I think there has to be a role for local zoning . . . You have to have some control over minimal things . . . [so there needs to be] three layers of protection.

Several interviewees suggested that many communities did not necessarily want to use zoning to veto shale gas development altogether but wanted to exercise some influence over the siting and conduct of drilling activities, and the distribution of economic benefits:

> I think that is really a kind of a democracy issue . . . I don't know that [people] want to stop it necessarily, but they want to have more control over it.
>
> (Benjamin)

> [E]specially when you look at the obscene profits those companies are making, we ought to be able to share some of that, whether it's financial or just the aesthetic and the use and quiet enjoyment of your property.
>
> (Christopher)

To that end, Kitze (2013) has proposed 'moving past preemption' in order to enable local communities 'to create and fulfill their own unique visions of how they will live' (2013: 395). She argues that local control of shale gas development should be enhanced through a more formal and meaningful role in state and federal regulatory processes. In a similar vein, interviewees for this research believed that increased participatory practices which enabled dialogue, and supported the inclusion and active involvement of all relevant stakeholders, would be of benefit:

> Some counties . . . have been more progressive than other counties in terms of . . . getting a broad spectrum of interests at the table at a local level . . . to talk through the issues and the problems . . . [I]f you had . . . an inclusive process . . . a collaborative process that informs what the national standard should be, maybe [local communities] will feel that they've skin in the game.
>
> (Martin)

> [We need] to set a structure in place where you don't have winners and losers, but you have participants in a partnership.
>
> (Frank)

> [We need] to get a really deliberative process going, to get really sound regulations in place.
>
> (Benjamin)

Sandra noted that it is critical to provide necessary information and resources to facilitate public participation:

> I think communication is more important than anything, knowing what they want, how they feel, what they need to know when they have questions . . . [you need to] hold workshops, educate them.

Interviewees indicated that the media, in particular, need to improve access to what Edward termed 'good, clean objective information' to enable informed civic discourse and debate. He believed that some outlets had been 'less than fair' in reporting on the costs and benefits of shale gas development. Ryan said he was 'used to getting misquoted and misrepresented in the media'. Sean also remarked:

> I'm ashamed and embarrassed about the surface treatment of these issues by most of the media. I don't feel that they've delved into the complexities, I think they've stirred passions, I think they looked for simple, sizzly story lines. Investigative reporting has been minimal and when it has been occurring . . . there are some who have produced reports which I don't feel are objective, and that frustrates me.

Interviewees also believed that greater recognition of local communities as having a legitimate interest in shale gas development was fundamental to improving community capabilities to participate in shale gas development decision making processes. For example, Margaret observed:

> I think there is some merit to people feeling heard, you know what I mean? Even if they don't view it as impacting the process, getting to vocalise the concern and feeling heard and knowing that the media can be there and it's [an] open process.

Margaret added that decision making structures need to be sufficiently broad to take account of the full spectrum of concerns that might be expressed:

> Because of the nuances of the law, people go off on tangents about things that aren't related to that set of laws [and the decision maker] says, 'Well that's unfortunately not our scope of interest'. I think that is the frustration for people is that they've got to be specific concerns that are addressed by each regulatory body.

Edward emphasised that adequate recognition includes ensuring that individuals and groups opposed to shale gas development are not unfairly dismissed as 'NIMBYs' or radicals by either government or industry. Edward quipped that his environmental advocacy organisation had been referred to as the 'Evil Empire' in industry publications: 'Yeah, right – I think Ronald Reagan, he talked about Russia and the Soviet Union as being the Evil Empire!'

As noted in Chapter 2, such disrespect constitutes a structural injustice which can discourage legitimate objection and participation in decision making processes. As one example of this, some interviewees cited the previous government practice of monitoring anti-Marcellus individuals and groups. In one leaked intelligence bulletin (provided to law enforcement agencies as well as private gas drilling companies), a public screening of *Gasland* and a council hearing on a shale gas development approval were identified as sites where environmental extremists and militants against shale gas development would be in attendance.[58] In another bulletin, the Gas Drilling Awareness Coalition (an environmental activist group) was labelled a 'potential terrorist threat'; they subsequently sued two state officials and the intelligence gathering firm for defamation, and the matter was later settled out of court (Cusick 2015a). Despite being cleared of wrongdoing and engaging at all times in lawful activities, representatives of the Coalition have noted that their listing as a potential terrorist threat has discouraged people from attending their rallies, and has had a 'chilling effect upon freedom of speech' (Cusick 2015a). Surveillance activities have continued elsewhere, with recent reports indicating that law enforcement agencies have been sharing intelligence with the gas industry about activists and protest activity (Cusick 2015b). Eaton and Kinchy (2016) show that a lack of political opportunity can restrict capabilities to engage in collective action against shale gas development, which may lead to the misleading conclusion that silence is indicative of consent.

Interviewees also considered that the ability of gas companies to fund comprehensive and wide-reaching public relations campaigns reduced the visibility of those opposed to shale gas development, enabling the government to 'overemphasise the great of companies and underemphasise the rights of people who are potentially impacted' (Margaret). According to interviewees, industry advertising has, in some cases, become conflated with factual evidence, making it difficult for citizens 'bombarded with information . . . to sort it [and] categorise it' (Christopher). Further, while the industry was able to prominently frame its benefits in the wider public discourse, some landholders who had experienced some of the negative impacts of shale gas development were gagged from talking to the media and third parties via non-disclosure agreements, creating a further public information gap.[59] Settlement agreements and contractual arrangements between landowners and gas companies are typically unfiled out-of-court affairs, which makes it difficult to track the frequency and nature of complaints, and limits the availability of information on the potential harms and risks of shale gas activity to the public (Fisher 2015). Interviewees considered that to minimise the disparity between citizens and industry, it is critical to have increased public access to settlement agreements involving environmental or public health, safety and welfare hazards.

In the eyes of the interviewees, citizen capabilities to participate in land use decision making processes for shale gas development could be greatly strengthened through improved participatory conditions and recognition of non-dominant interests by powerful stakeholders. Combining these measures with local zoning powers that are underpinned by robust state and federal regulations was seen as critical to alleviating what many interviewees perceived as a situation of 'haves' and 'have nots', where 'the deck is stacked against the average person' (Christopher). As Pacheco (2015) notes, there are well-established models of cooperative federalism between the state and federal governments under major US environmental statutes, and these could be drawn upon to craft more collaborative and inclusive governance structures that promote environmental justice in shale gas development decision making.

Conclusion

Governor Tom Corbett declared, in his 2011 budget address, that he wanted to make Pennsylvania 'the Texas of the natural gas boom' (Corbett 2011: 6) and there is little doubt that his administration did all it could to achieve this aim, most controversially with the introduction of *Act 13*. In the eyes of the interviewees for this case study, *Act 13* represented environmental injustice in terms of its content and formulation. Its express command that shale gas development be allowed in all zoning districts failed to recognise the interests of local communities, and its limited opportunities for public participation restricted the capabilities of citizens to control the impacts of development. The processes for developing this legislation from the outset were considered to have been heavily influenced by industry, and further limiting of citizen agency. Like the previous case studies, the Marcellus Shale case has demonstrated how injustice may be experienced by stakeholders at all stages of the land use decision making process – including at the early phases of developing and implementing the relevant regulatory framework.

The Marcellus Shale case study re-emphasises the fundamental link between procedural fairness and recognition as well as the overarching nature of capabilities. Those concerned about the impacts of shale gas development had little opportunity within the parameters of *Act 13* to object, meaning that their interests were not sufficiently recognised within the regulatory framework. Even when participatory mechanisms were present – such as in the DEP's permit approval processes – circumstantial factors of those who would seek to challenge development, such as rural communities, created barriers to participation. By fashioning a narrow structure for participation and not accounting for specific impediments to participation (e.g. a lack of resources to review permit applications within the allotted time frame), the government obscured the interests and concerns of many stakeholders within the regulatory framework. *Act 13* clearly constrained the capabilities of individuals and communities to manage the local impacts of shale gas development, leading to the view that government and industry 'just don't care':

> [T]hey're operating from a very different view of what they want to achieve and how they're going to achieve it, and they've been running people over and that's part of the problem.
>
> (Matthew)

Like the Bulga and Namoi catchment case studies, central to the perceived lack of agency on the part of local communities and individual citizens was the interplay between power dynamics and scale framing. Specifically, the government had *power over* the governance process, flowing from its 'formal and acknowledged' role as organiser (Van Lieshout et al. 2014: 5). It was able to 'set the agenda, steer towards particular outcomes (including the reframing of certain scale frames), control the information provided, and decide on note-taking, conclusions and documentation' (Van Lieshout et al. 2014: 5). Meanwhile, the shale gas industry and its supporters had *power in* its interactions with government. The ability for industry to fund wide-reaching public relations campaigns about the benefits of shale gas development provided a significant discursive capacity to frame the scale of benefit at the broader state and national level (Sica 2015; Andrews and McCarthy 2014; Van Lieshout et al. 2014); while considerable political contributions reinforced industry status and provided additional leverage in relations with government. By comparison, citizen and community actors had little in the way of power in their interactions within governance episodes (e.g. the Marcellus Shale Advisory Commission), and they had limited *power of influence* over governance processes (Van Lieshout et al. 2014).

Although the balance of power swung back towards those concerned about extractive development in the wake of the Supreme Court's decision in *Robinson Township*, the positions of the key actors are still far from even. Procedurally, the precedent value of the application of the *Environmental Rights Amendment* is still uncertain and, pragmatically, the return of local government zoning power has reiterated concerns about local capacity to engage effectively in these practices. There is a strong argument in favour of local control but the reflections of the interviewees in the Marcellus Shale case study illustrate how the quality of participation can still be eroded by a lack of resources and local negotiations dominated by industry influence.

Governor Tom Corbett was defeated at the 2014 gubernatorial election by Democrat Tom Wolf, who campaigned against Corbett's management of shale gas development in the state. Shortly after taking office in early 2015, Governor Wolf imposed a moratorium on new drilling leases in state parks, expressed strong support for a severance tax on production and promised to finalise environmental protection regulations for gas well sites (Finley 2015; Clough and Bell 2016). It remains to be seen how, under this new leadership, the state government will deal with the aftermath of the *Robinson Township* litigation and, in particular, whether there will be concrete efforts to enhance community capabilities to participate in shale gas development decision making.

In 2016, there were some moves towards increasing public participation in oil and gas siting decisions in Pennsylvania. Specifically, the DEP recently added oil and gas permits to the list of applications that trigger enhanced public participation procedures in 'environmental justice areas' (Department of Environmental Protection n.d.; Hopey 2016). Environmental justice areas are census tracts where more than 20 per cent of individuals live in poverty, and/or 30 per cent or more are in a minority group. However, recent demographic data show that, in the census tracts throughout Pennsylvania where oil and gas activity is occurring, the population is on average 98 per cent white, and less than 20 per cent of residents are considered to be living in poverty according to federal poverty guidelines (Jalbert 2016). With projected development likely to take place in similar demographic areas, it is unlikely that many future siting decisions will be subject to the DEP's enhanced public participation procedures. For this reason, Jalbert (2016) argues that the DEP's 20/30 rule 'misses the mark' because its threshold for disadvantage does not recognise the realities of economic and social vulnerability in rural areas. Many census tracts where shale gas development is occurring contain a significant percentage of individuals defined as 'working poor' (those earning less than two times the poverty line – currently US$23,760 per annum) (Jalbert 2016). Further, a substantial amount of gas development is taking place in census tracts where the median house price is below US$200,000 (Jalbert 2016). Thus, there are citizens 'on the margins of vulnerability' (Jalbert 2016) who may be negatively impacted by shale gas development, yet do not have access to appropriate public participation opportunities to voice their concerns. This underscores the need for a more comprehensive and nuanced understanding of vulnerability, and a multifaceted definition of environmental justice which seeks to extend 'the levers of public input . . . to environmental justice communities and beyond' (Jalbert 2016).

Following the Supreme Court's latest findings in *Robinson Township* there will, undoubtedly, be further debate over local control of shale gas development. However, the *Robinson Township* saga need not be read as a complete rejection of state regulation of shale gas development. Indeed, as Fershee argued following the original plurality decision, despite the 'lofty rhetoric' expressed by the Court it may simply be the case that the *Environmental Rights Amendment* contained in Article 1, §27 of the *Constitution* requires citizens and local governments to have greater participation in state-level shale gas decision making and that *Act 13* failed to provide this (2014: 857). Pacheco concurs, noting that *Robinson Township* is 'best understood as a call for a fairer process that offers more meaningful opportunities for local governments to influence fracking regulations' (2015: 392). As the interviewees for this research have indicated, there are several ways in which the capabilities of communities to participate in shale gas governance could be enhanced. The legacy of the *Robinson Township* litigation, then, may be that it serves as a catalyst for the state government to consider more deeply how the capabilities of communities to participate effectively in shale gas development decisions can be strengthened.

Notes

1 As Anderson explains, mineral rights 'can be severed in two ways: by deed or via an oil and gas lease ... An oil and gas lease is not like a typical "lease", but is real property interest, a fee simple determinable in the oil and gas with a reversionary interest in the lessor, plus an implied easement to access the surface estate for purpose of developing the oil and gas estate' (2013: 138–139, references omitted).

2 *Gasland* (2010), New Video Group, available at www.gaslandthemovie.com. Now infamous for its depiction of a Colorado resident lighting water from a kitchen faucet

(asserted by the landholder to be a result of methane contamination of groundwater sources caused by drilling), the film focused on experiences of landholders with natural gas drilling on or nearby their properties.

3 *Truthland* (2012), Energy In Depth and Independent Petroleum Association of America, available at www.truthlandmovie.com/

4 Local government in Pennsylvania is separated into 67 counties, which are then subdivided into municipalities comprised of various cities, boroughs and townships. County government functions vary depending upon population size and the nature of services required, but can include welfare services, corrections and law enforcement, planning and economic development; while municipal governments are responsible for public safety (e.g. police and fire services), subdivision and land development.

5 *Oil and Gas Act 1984*, P.S. §601.102, repealed by 2012 *Pa. Legis. Serv. Act* 2012–13 (H.B. 1950).

6 964 A.2d 869 (Pa. 2009).

7 964 A.2d 855 (Pa. 2009).

8 Similarly, in *Range Resources Appalachia, LLC v. Blaine Township*, 2009 WL 3515845, a disclosure ordinance which provided the Township with the ability to prevent drilling where a gas company had three or more violations of the law was seen to be pre-empted by the *Oil and Gas Act* because it provided the Township with an 'almost unbridled discretion to deny permission to drill'.

9 58 Pa. Cons. Stat. §§2301–3504 (2012).

10 Ibid. §§3302, 3303; see also Marcellus Shale Advisory Commission (2011: 65).

11 Ibid. §3305(b).

12 Ibid. §3211.

13 Ibid. §3212.1(a).

14 Ibid. §3212.1(b). Municipalities are also unable to appeal permit decisions: Ibid. §3212.1(d).

15 Ibid. §3212.

16 Ibid. §3211.

17 *Robinson Township v. Commonwealth of Pennsylvania*, 52 A.3d 463 (Pa. Commw. Ct. 2012).

18 Ibid. at 484.

19 Ibid. at 481–2.

20 Ibid. at 496.

21 Ibid. at 495.

22 Ibid. at 484.

23 Ibid. at 490.

24 Ibid. at 493.

25 *Payne v. Kassab* 312 A.2d 86 (Pa. Commw. Ct. 1973).

26 *Robinson Township v. Commonwealth of Pennsylvania*, 52 A.3d 463 at 489 (Pa. Commw. Ct. 2012).

27 *Robinson Township v. Commonwealth of Pennsylvania*, 83 A.3d 901 (Pa. 2013).

28 A plurality opinion is an opinion from several judges which, while not receiving the support of more than half of the judges, receives more support than any other opinion rendered. A total of six Supreme Court justices heard the *Robinson Township* appeal.

29 Ibid. at 981–2.

30 Ibid. at 1006–7: 'once a state authorizes political subdivisions to zone for the best interests of the health, safety and character of their communities, . . . and zoning ordinances are enacted and relied upon by the residents of a community, the state may not alter or invalidate those ordinances, given their constitutional underpinning. This is so even if the state seeks their invalidation with the compelling justification of improving its economic development' (per Justice Baer, references omitted).

31 *Robinson Township v. Commonwealth of Pennsylvania*, 83 A.3d 901 at 957–8 (Pa. 2013).

32 Ibid. at 945, 960.

33 Ibid. at 954.
34 Ibid. at 953.
35 Ibid. at 985.
36 Ibid. at 980.
37 Ibid. at 979.
38 Ibid. at 983–4.
39 Ibid. at 984.
40 Ibid. at 964.
41 Ibid. at 980–1.
42 Ibid. at 976.
43 Ibid. at 1010. The plurality noted in this regard that the Commonwealth 'should be aware of and attempt to compensate for the inevitable bias toward present consumption of public resources by the current generation, reinforced by a political process characterized by limited terms of office' (Ibid. at 959). See Daly and May (2015: 158) for further discussion of this point.
44 Ibid. at 1011.
45 Ibid. at 1013. See also Fershee (2014: 849–850), who argues that the plurality substituted a perception of the risks of fracking for factual evidence and research: 'Drawing conclusions based on the court's "sense" or perception of the facts, rather than a review of relevant research . . . is bound to produce bad outcomes.'
46 Ibid. at 1010.
47 Ibid. at 1015.
48 *Robinson Township v. Commonwealth of Pennsylvania*, 96 A.3d 1104 (Pa. Commw. Ct. 2014).
49 Ibid. at 1120–22.
50 Ibid. at 1120.
51 *Range Resources Appalachia LLC v Salem Township*, 964 A.2d 869 (Pa. 2009).
52 E.g. *Pennsylvania Environmental Defense Foundation v. Commonwealth of Pennsylvania*, 108 A.3d 140 at 156 n. 37 (Pa. Commw. Ct. 2015) and *Kretschmann Farm, LLC v. Township of New Sewickley*, 131 A.3d 1044 at 1059 n. 20 (Pa. Commw. Ct. 2016). See Hagen-Frederiksen 2016: 385–388. An appeal against the decision in *Pennsylvania Environmental Defense Foundation* has recently been heard in the Supreme Court, seeking to affirm the meaning of the *Environmental Rights Amendment* in Article 1, §27. At the time of writing, a decision had not been rendered. See: *Pennsylvania Environmental Defense Foundation v. Commonwealth of Pennsylvania et al.*, 10 MAP 2015 (Pa. Sup. Ct.).
53 *Robinson Township v. Commonwealth of Pennsylvania*, 96 A.3d 1104 (Pa. Commw. Ct. 2014). Section 3241(a) was upheld because it did not amount to the taking of property for private purposes, as gas transportation is considered a public service.
54 66 Pa. Cons. Stat. §1104.
55 See, e.g., *Sunoco Pipeline, L.P. v. Loper*, 2013-SU-4518–05 (C.P. York, February 24, 2014); *Sunoco Pipeline, L.P.*, No. P-2014–2411941, 2014 WL 5810345; *In re Condemnation of Sunoco Pipeline, L.P.*, 2016 Pa. Commw. Ct. LEXIS 326 (Pa. Commw. Ct. July 14, 2016).
56 *Robinson Township. v. Commonwealth*, No. 104 MAP 2014 (Pa. Sept. 28, 2016).
57 Citing brief of Amicus Curiae Pennsylvania State Association of Township Supervisors at 3, *Robinson Township v. Commonwealth of Pennsylvania* (No. 63 MAP 2012).
58 A copy of one of the briefings is available online at: https://www.propublica.org/documents/item/pennsylvania-intelligence-bulletin-no.-131-aug.-30-2010.
59 See e.g. *Hallowich v. Range Resources Corp* 30 Pa. D. & C. 5th 91 (March 20, 2013).

7 The search for justice in the governance of extractive resource development

Introduction

In several countries, the expansion of extractive industries has increased conflict over land use, generating a critical policy concern for various levels of government. Such contests are, of course, not novel; resource development inevitably brings about change, and there have always been trade-offs between economic, environmental and social interests. However, the current intensification of extractive development, combined with new and potentially risky extractive technologies, has created different challenges. As illustrated in the case studies described in this book, governments have attempted various strategies to resolve conflicts accompanying developments, but their responses have had limited success and, in some cases, have created increased costs and other consequences for communities.

Using the concept of environmental justice to guide discussion, this book focused on the impact of regulatory and policy mechanisms in the management of competing land use interests, exploring why these interventions have failed to meaningfully transform situations of conflict. The case studies revealed that, for the most part, the governance mechanisms implemented did little to address the primary concerns of many actors in disputes over extractive development. In addition, the case studies exposed a range of systemic factors that frequently privileged industry interests and marginalised alternative values and concerns. This resulted in claims of injustice that spanned not only the outcomes of decisions, but also how decision making processes were conducted, and how individual and community well-being was impacted as a result.

Chapters 1 and 2 scanned the vast theoretical literature on the concept of environmental justice. Many scholars have discussed, from a theory perspective, the existence of injustice and how it should be remedied but there is a limited body of research which empirically applies justice concepts to the practice of land use decision making, and there is even less which explains the perceptions of injustice which drive land use conflict over extractive resource development. This book makes a unique contribution by presenting a careful and detailed analysis of these factors within the context of extractive development (Chapters 3 to 6), providing critical insight into why such conflicts elude resolution. The purpose of this final chapter is to bring together the findings from the case study chapters, situating injustice as a principal driver of land use conflict. It identifies the key environmental justice themes that emerged from the case studies; specifically, the capacity of stakeholders to participate in decision making, and the framing of issues that downgrades the importance of place-based claims. The concluding parts of this chapter suggest reforms to the governance of extractive resource development that would ameliorate conflict.

Overall, the case studies indicate that injustice is a complex systemic issue rather than the result of the isolated operation of discrete instruments or institutions. Injustice is produced through the interactions between various actors and numerous intertwined structural processes. To resolve conflict, decision makers need to be prepared to engage more deeply with the concept of justice at a broad 'meta-governance' level. In short, to ensure whole-of-system integrity, decision makers need to embed justice as a guiding principle for governance.

Environmental injustice and land use conflict

Environmental justice, as defined in this book, integrates distribution, participation and recognition with the notion of 'capabilities', recognising the interrelationship between different dimensions of justice. Applying the concept of capabilities to cases of conflict identifies whether social arrangements foster the freedom and opportunities necessary for people to achieve desired functionings and well-being, considering whether their choice and agency are enabled or constrained by multiple overlapping political, social, cultural and institutional factors.

A focus on capabilities highlights 'what matters' (Davoudi and Brooks 2014: 2689) in claims for justice, how inequality is experienced through what people have the freedom to do and be, and why it is considered wrong in terms of the functionings they value. Environmental injustice can, therefore, be located in less-obvious situations – not only those concerning disadvantaged and minority groups – where 'unequal become[s] unfair' (Davoudi and Brooks 2014: 2686) for a range of stakeholders through the interaction between power and place. A focus on capabilities takes the analysis of injustice beyond a discussion of normative justice theories and empirical accounts of inequality towards an understanding of process: *why* things are how they are. This overcomes the 'simple judgment that anything unequal is also wrong' (Walker 2012: 217), recognising that justice (and injustice) is situated and contested.

By examining the case studies (Chapters 3 to 6) through the lens of capabilities, I was able to identify which freedoms or opportunities decision makers and decision making processes were impacting. I focused particularly on unravelling the ways in which various actors and institutions interacted with one another, and how power manifested itself in these relationships to either enable or constrain capabilities. Two capabilities emerged as particularly relevant to the stakeholders I interviewed: the ability to participate effectively in land use decision making, and the ability to maintain place attachment.

Effective participation in land use decision making

Many stakeholders were unhappy with the nature of public participation processes (procedural justice), with how those processes were conducted, and with how they had been treated (recognition). Stakeholders maintained that their capability to effectively participate in land use decision making was limited in a number of ways across the entire decision making process, including the formulation of the decision making processes. Where opportunities to participate were non-existent or minimal, stakeholders perceived that capabilities had been constrained. For example, in Part 2 of the Namoi catchment case study, interviewees considered that there were limited opportunities to participate in decisions about CSG exploration; and in the Marcellus Shale case study, interviewees believed that the implementation of *Act 13* removed opportunities for community participation in development decisions by the restriction of local zoning powers.

Even in cases where participatory mechanisms existed, the cases show that *effective* participation was still found wanting. In a number of instances public participation in land use decision making was limited to written submissions or once-off public hearings, which did not permit the multi-directional consultation that individuals and communities desired. Interviewees complained that information about proposed developments was received too late and did not reflect the concerns expressed by stakeholders. Even when stakeholders used FOI and government disclosure legislation (potentially providing limited access to government-held information only) to request access to information (such as in the Bulga case, and in Part 2 of the Namoi case study concerning CSG exploration in The Pilliga), proponents were able to resist providing information by invoking confidentiality.

For many stakeholders interviewed, participation in decision making processes was limited to responding to proponent claims about economic benefit and scientific risk through written or verbal submissions. This required them to review a significant volume of technical data, usually with limited time. Many interviewees cited the difficulties that this posed, particularly the personal costs to participants, and the impact it had on the quality of their responses. In some cases, stakeholders needed to rely on expert testimony to challenge proponent claims but the cost and time needed to do so hampered these efforts. The impact of other structural factors exacerbated these disadvantages for certain groups. For example, while the electronic distribution of information about a development project can enhance its reach, it might not be available to those who do not have access to a computer or who are not computer literate, such as older citizens (as noted in the Bulga case study) or those from a minority cultural group (like the Gomeroi people in Part 1 of the Namoi case study). In some cases, rurality compounded barriers to participation, particularly where attendance at public meetings or hearings required travel to other places.

The evidence indicates a great inequity in public participation mechanisms that set well-funded corporate entities against underfunded or even non-funded community and interest groups, who may have to balance other family and work responsibilities alongside their participation in land use decision making processes (Radow 2010). As Richardson and Razzaque note:

> Excessively technical and bureaucratic procedures for public involvement can be a major hurdle for fruitful consultation. Complex, encyclopaedic-size EIA [Environmental Impact Assessment] reports, for instance, tend to hinder rather than facilitate public scrutiny of proposed developments. Lack of technical support and difficulties in gaining access to clear information can diminish the public's ability to provide a meaningful voice in decisions.
>
> (Richardson and Razzaque 2006: 193)

But development proponents not only possessed advantages in time, technical knowledge and economic resources for participation in land use decision making, they also enjoyed significant political influence, which they could leverage to shape the parameters and conduct of the assessment processes. In the Marcellus Shale case study, for example, the composition of the Marcellus Shale Advisory Commission for shale gas regulation was heavily weighted towards industry interests. Similarly, in the Bulga case study, stakeholders believed that the swift amendments to the *Mining SEPP* were prompted by requests from the proponent to override the Land and Environment Court's decision to deny the expansion of the Mount Thorley–Warkworth coal mine. In the Namoi catchment case study, despite broad public consultation in the development of the SRLUP, stakeholders believed that industry lobbying diluted the legislative provisions, providing fewer safeguards for agricultural land. In each of these cases, the resulting governance mechanisms seemed to favour industry, failing to recognise non-dominant interests.

The power of proponents and industry groups to frame the scale of conflict was pivotal to their political influence. As noted in Chapter 2, scales – local, regional or national – have a material component, but they are also discursively shaped by social and political relations. Across the case studies, notions of the 'public interest' being concerned with regional and state scales of economic and energy security displaced the local environmental and social scale. The significant capacity of proponents and industry to fund wide-reaching advertising campaigns, and their access to politicians for lobbying, enabled the construction of a persuasive – and pervasive – narrative that positioned extractive development as 'natural, normal and legitimate as possible' (González 2006: 838). Existing 'scalar structures' (MacKinnon 2011) – notably, historical constructions of the benefits of extractive development to struggling rural areas – lent further support to industry influence. This provided proponents and industry with power in their interactions with government to encourage particular governance outcomes. Further, in several cases, the socio-spatial characteristics of rural areas – particularly low population density – limited the ability of stakeholders to publicise their concerns at the wider scale, further obscuring their perspectives.

It is widely recognised that extractive industries engage in self-interested lobbying; however, what is less acknowledged is that these campaigns have a significant capacity to restrict discourse and influence public sentiment as well as government responses: '[P]owerful actors' use of "framing" and "muting" to suppress dissent can have long-lasting, harmful ramifications and may be difficult for those participating in the debate to perceive' (Eisenberg 2015: 205).

Scale framing efforts may be employed to 'corral' public opinion and manufacture apparent consent for governance interventions that support extractive activities (Hudgins and Poole 2014: 305). In the case studies, governments harnessed industry discourse to justify policy and regulatory reforms, and used their authoritative power over governance processes to establish the parameters of conflict resolution (Van Lieshout et al. 2014). The claimed economic and energy security benefits of resource extraction provided an ethical rationale of the 'public good' to govern in ways that facilitated development. Governments framed concerns about social and environmental impacts within broader capital interests (particularly in the case of unconventional natural gas, which was presented as a cleaner source of energy than other fossil fuels, and a bridge to transition to renewable energy). Decision making mechanisms became aligned with the imperatives of development at the broader economic and energy security scale, normalising and de-politicising 'technocratic' expert opinions in favour of the market, and conflating industry interests with the public interest (Mercer, de Rijke and Dressler 2014: 284). The processes often involved the 'fetishization of the scientific process' (Simonelli 2014: 274), where expert opinions were positioned as 'objective' and 'apolitical' (Richardson and Razzaque 2006: 171), marginalising other forms of knowledge, including place-based perspectives.

In the case studies, those seeking to challenge development were often required to engage 'authoritative' scientific opinions and methodologies, or risk their concerns being discredited and dismissed. In many instances decision making processes excluded or gave little weight to non-expert perspectives, such as citizen concerns about place-based impacts and the complexity of social and cultural concerns. Generally, such issues needed to be 'transmuted into the restrictive epistemology of quantitative social economic modelling' (Martin and Kennedy 2016: 105). In many instances, the conflict to be decided was largely reduced to a narrow question of negotiating the appropriate trade-off between competing economic and enviro-social interests. The narrative of balancing competing interests was an important discursive tool for governments to shield themselves from criticisms of unfairness, enabling them to be positioned as an independent arbiter of disputes. The limited capacity to contest development decisions – evident in the Bulga case and in Part 1 of the Namoi catchment case study, where bureaucratic decisions to hold 'public hearings' closed off opportunities for merits review in the Land and Environment Court – effectively removed issues from legal contestation. These factors not only constrained the capabilities of individuals and communities to participate effectively in decision making, but also impaired individual and group capabilities to maintain place attachment and exercise autonomy over their local environment – a point which will be explored further in the following subsection.

Other actions, such as bureaucratic decisions to hold 'public hearings' to close off opportunities for merits review in the Land and Environment Court, the removal of funding for public interest legal services, attempts to limit standing and curtail protests, and the derogatory characterisations of individuals and groups opposed to development and the dismissal of their concerns as self-interested, reinforced a lack of recognition of alternative interests and concerns. In effect the 'displacement, disempowerment and ignorance' of alternative perspectives resulted in the local scale becoming a source of environmental injustice (Jessup 2013: 106).

Despite apparent opportunities for individuals and communities to be involved in decision making processes, these avenues offered 'no more than a thin patina of inclusion' (Jarrell, Ozymy and McGurrin 2013: 184). This is a common complaint about mechanisms for public participation in environmental decision making; the existence of opportunities for participation may not necessarily translate into *effective* participation, due to financial, technical and other structural barriers that impair input. Ineffective participation undermines the credibility of decision making frameworks, and their capacity to neutralise social conflict. Jarrell et al. observe:

> Failure to be inclusive and transparent in what issues reach the agenda and not properly weighing alternatives (including citizen proposals) may render perceptions of the process as illegitimate. Citizens are likely to feel that public officials are engaging in selective political agenda setting that ultimately tears down opportunities for collaboration.
>
> (Jarrell et al. 2013: 186)

The case studies in this book have demonstrated that restricted capabilities to participate effectively in land use decision making can cultivate perceptions of injustice and foster opposition. As Davoudi and Brooks note, 'being subjected to the harms in which we have no choice or responsibility reinforces people's sense of injustice' (2014: 2690). Even if concerned stakeholders eventually achieve just outcomes – such as with the buy-back of the Caroona coal exploration licence, in Part 1 of the Namoi catchment case study, and in the Marcellus Shale case study, with the Supreme Court decision to deem parts of *Act 13* invalid – they will perceive the outcomes as contingent upon non-processual factors outside their control (Sharma-Wallace 2013). Prior processes can continue to be viewed as a source of injustice, which can have ongoing ramifications for the trust and legitimacy of government – even in unrelated and future decisions (Gross 2014).

The corollary is that when those impacted perceive processes to be just and conducted in a fair and impartial manner from the outset, decisions are more likely to be accepted, regardless of substantive outcomes (Syme and Nancarrow 2012; Clayton et al. 2013; Gross 2014). To overcome perceptions of injustice and minimise conflict it is fundamental that decision makers understand how capabilities of effective participation may be impaired – including the nature and availability of participatory opportunities, and the structural factors which impose barriers to participation – and remedy these constraints.

Maintaining place attachment

As noted in the previous subsection, the capabilities of individuals and groups to participate effectively in land use decision making was often constrained because of the restriction or removal of certain factors for consideration from decision making processes. In particular, the privileging of expert scientific evidence and the limited opportunities to raise place-based concerns diminished the capacity of citizens to contest industry and government hegemony. But, in addition to impairing participatory capabilities, these factors also impacted place attachment. That is, governance arrangements restricted the capability of individuals and communities to maintain their social and emotional connections to place, and their autonomy to protect the landscapes and livelihoods that they valued.

As discussed in Chapter 2, place attachments are the multidimensional, dynamic bonds formed through interactions with physical spaces, and the groups and individuals within them. Places are imbued with diverse and overlapping meanings; landscapes are not only physical spaces but also reflect a 'relationship and a process' (Willow, Zak, Vilaplana and Sheeley 2014: 57). It is these relationships, and 'not just place qua place, to which people are attached' (Low and Altman 1992: 7). Attachment to place creates a 'secure intersubjective space' that operates as support for one's way of life and, from this base, more intricate and integrated forms of agency can develop – including interpersonal relationships, relationships with non-humans and objects, and social institutions (Groves 2015: 855–856): '[Place] attachment weaves together place, identity and agency in helping to domesticate, for individuals and the collectivities of which they are part, an intrinsically uncertain future' (Groves 2015: 855).

The fixed location of mineral and gas resources means that extractive development typically results in the significant transformation of particular places (Van Wagner 2016a). These impacts can have a tangible impact on ecological dependence and emotional well-being. This is, of course, not exclusive to rural areas but, in many ways, the physical landscape is often deeply tied to the identity and social characteristics of rural residents. As Bartel and Graham note, notions of place may be 'particularly strong for the identity of farmers, which is linked to geographical and social distance from urban political centres and services, and also hence including self-reliance and independence' (2016: 276). For different reasons, the physical landscape is also central to the identity of Indigenous peoples in rural areas, whose cultural identity is fundamentally connected to their sense of place (Rose 1996). Changes to the characteristics of places can 'cease to sustain those who inhabit them, as the "secure space" that makes embedded agency possible decays' (Groves 2015: 857), destabilising individual and collective identity.

A failure to support the agency of individuals and groups to influence their future through decision making can disrupt place attachment (Groves 2015).

When those who challenge development are required to do so according to the terms and language of those in power they are forced to obscure their place-based values and are denied their capacity to manage place transition. As Groves explains, land use governance mechanisms typically reward those actors (such as proponents of development) who are 'able to marshal the requisite expertise in risk analysis', while at the same time minimising the agency of others, which consequently 'crowds out the futures they, for their part, envision and anticipate' (2015: 869).

In the case studies, a concern of many of the stakeholders interviewed was the concentrated local environmental harm that had already resulted from extractive development, or would likely eventuate, and the inability to manage the uncertainties of place disruption. In the Bulga case study, changes to the composition of the community that would follow the mine's expansion, increased noise and dust impacts, and the potential for significant biodiversity loss compounded the loss of sense of place being felt by residents living with the impacts of existing operations. The community's thwarted attempts to raise these concerns resulted in feelings of dispossession and disempowerment, and a loss of control over their future. In the case of proposed coal mining on the Liverpool Plains in the Namoi catchment, interviewees were concerned not only about the potential physical impacts upon the landscape and agricultural and cultural activities, but also with the threats to livelihood and identity for farmers, and Indigenous peoples. Farmer participants in this research linked their connection to the land with a long history of stewardship and land management practices embedded in their knowledge of place, and deplored the loss of control over their future. Indigenous people feared that the removal of the 'grinding grooves' sites would impact their ability to engage in cultural practices, which would undermine their identity and future. Stakeholders viewed CSG exploration in The Pilliga as a threat to biodiversity as well as the water resources in the Great Artesian Basin, raising questions about the interspecies and intergenerational future of shared public spaces, and about who has the right to use and maintain them. In the Marcellus Shale case study, interviewees described frustrations with their reduced capabilities to control the rapid development of shale gas, and their lack of agency over environmental sustainability, human health, community continuity and preservation of the rural way of life.

Irrespective of whether the impacts of place transition are 'proven, perceived, or anticipated' (Willow 2016: 771), the inability to deal with uncertainty can have a distinct detrimental effect upon individual and community identity and well-being. Such harms, as Groves notes, are not just indicative of a failure to acknowledge 'difference' – in other words, misrecognition – but restrict capabilities to navigate the future:

> Witnesses to such losses are left scrambling to reform communal and individual strategies for dealing with uncertainty . . . Injustice is made tangible in such cases via descriptions given by people of the erosion of their sense of themselves as doers and actors.
>
> (Groves 2015: 858)

Interviewees across the case studies described how their capabilities for dealing with feared and actual place disruption had been constrained by the inability to effectively express place-based concerns. Residents in the Bulga case study, concerned that their village would 'die' as a result of the Mount Thorley–Warkworth coal mine expansion, felt that they were oppressed and dominated within land use decision making processes, and unable to effectively manage place transition. Similar sentiments were expressed by interviewees in the Namoi catchment and Marcellus Shale case studies. By being unable to articulate effectively what matters within land use decision making processes, identity and agency becomes unstable and incapable of withstanding change (Groves 2015: 869). This, in turn, can lead to place-protective behaviours, as well as social conflict (Devine-Wright 2011a: 337; Sherval and Hardiman 2014; Bartel and Graham 2016).

The case studies in this book have demonstrated that an erosion of place attachment can result from land use decision making processes which do not acknowledge that the 'connection between people and land is more than simply biological and economic' (Sherval and Hardiman 2014: 180). Interviewees articulated landscapes as 'places', with specific visions for the management and allocation of natural resources; yet, the capability to realise these visions was constrained by governance arrangements that marginalised place-based concerns. Social conflict and regulatory failure can result where there is a 'lack of recognition of place-specifics by government and laws that impose universal requirements' (Bartel 2014: 891), driven by a sense of injustice that arises from the disempowerment of individuals and groups to control and protect their surrounding environment (Willow et al. 2014). Greater attention to place-based concerns within governance arrangements is critical to improving the ability to maintain place attachment, and overcoming perceptions of injustice.

From a capabilities perspective, the governance mechanisms employed in each of the case studies considered in this book illustrate 'a restrictive and insufficient notion of justice' (Walker 2012: 52). While ostensibly implemented to deal with conflict over land use, the governance interventions failed to reduce inequalities and, in some instances, appear to have exacerbated and even produced other disadvantages. Apparent across the case studies was an unequal distribution of benefits and environmental harms and risks, uneven participatory opportunities in decision making, and a lack of recognition of non-dominant interests, which, overall, restricted the capabilities of certain stakeholders to realise valued functionings. Governance arrangements constrained the ability to participate effectively in land use decision making, and the ability to maintain place attachment.

While most of the communities examined in the case studies would not generally be considered as conventionally marginalised people, injustice was nonetheless found in the disenfranchisement generated by land use decision making frameworks. Environmental injustice arose because the capabilities of self-identity and self-determination were impaired by the limited participatory opportunities afforded to communities, and the exclusion of their non-dominant interests. These were direct consequences of the disparate power relations between rural communities, extractive industries and governments. The structural, material and discursive capacities of industry and government shaped the governance system, influencing who controlled – and who was excluded from – land use decision making. Curing environmental injustice and overcoming the resulting conflict requires engaging with principles of justice, and confronting power asymmetries within governance systems.

Ensuring just decision making

It is often assumed that conflict over extractive development arises because the interests of proponents of development, and those who stand to be impacted by it, diverge. That is, of course, part of the picture, but it oversimplifies the complex dynamics at play. The case studies detailed in this book demonstrate that conflict also arises from perceptions of injustice in the processes and interactions relevant to decision making concerning development. This is particularly so if the processes and interactions fail to engage meaningfully with what matters to stakeholders. Governance arrangements which marginalise or disregard concerns, or which fail to deal with the central issue of power dynamics are unlikely to shift situations of conflict towards more socially cohesive outcomes. By prioritising particular interests, governance arrangements can become static and maladaptive to changing values and conditions. This reduces their capacity to respond to new complexities and to manage unanticipated disputes (Martin and Kennedy 2016).

As a result of the findings from the case studies discussed in this book, two key areas for governance reform emerge as particularly relevant for reducing conflict: enhancing public participation in land use decision making and incorporating place-based knowledge.

Enhancing public participation

Public participation is widely accepted as an essential component of good environmental decision making. The literature discussed in Chapter 2 highlights the conditions for more effective participation, including: clear and early notice of issues requiring decisions that will impact the environment; provision of relevant information; multi-directional consultation; meaningful input into decision making; and access to review functions. Many studies have identified that such provisions can produce better informed decisions and encourage acceptance of outcomes (Hunold and Young 1998; Smith and McDonough 2001); but the case studies demonstrate that a failure to improve the capabilities of individuals and groups to engage, and an ignorance of power imbalances can render participation ineffective. Such failure can lead to a lack of trust in decision makers, rejection of decisions and conflict.

Proponents and governments can overcome certain financial, technical and other structural barriers to participation by attending to the manner and methods of participation (e.g. allowing longer time for submissions, or simplifying technical data), and recognising and potentially offsetting the financial costs of participation. However, to achieve the deeper engagement that would build the overall capabilities of citizens to participate effectively in land use decision making, communities need to be assisted to address their own self-defined needs (Measham 2007: 146). At a minimum, capacity-building should involve training to develop communication and negotiation skills, and to provide the expertise to understand government processes and technical information (Conrad et al. 2011). Such initiatives should also focus on supporting communities to communicate their own place-based knowledge in a two-way learning exercise (Andrew and Robottom 2005; Reed 2008).

Importantly, stakeholders need to have the ability to influence decisions (Reed 2008). Power imbalances must be confronted – if 'instrumental logic automatically precludes the advancement of procedural or interactional forms of justice' (Kemp et al. 2011: 103), then participants will continue to feel as though they have little voice. In the case studies most participatory processes did not satisfy the level of public participation desired by communities, and afforded little opportunity to influence outcomes. That is, they were akin to the lower rungs on Arnstein's (1969) 'ladder of citizen participation'. If a decision is not likely to be influenced by participants, then participatory processes will be perceived as 'window dressing' (Cotton 2016: 11), leading to frustrations and potential conflict. Where conflicts of interest including agenda management are apparent, and privileged groups are able to unduly influence outcomes, then decisions will not be seen as legitimate (Jarrell et al. 2013; Cook 2015). Opportunities to participate must translate into inputs that are meaningfully considered by decision makers, and seen to be integrated into decisions (Richardson and Razzaque 2006; Radow 2010; Conrad et al. 2011; Preston 2015a). Thus, policy and regulatory requirements to consult stakeholders or establish participatory processes should provide clear standards for public participation; they should detail how participation will occur, and how the inputs from participation will be used and evaluated.

Policy and regulatory mechanisms should also include a requirement to assess the effectiveness of participatory processes; without monitoring and evaluation, decision makers may hear, but not listen to, public views (Conrad et al. 2011). Richardson and Razzaque note the need to shift the onus of ensuring effective public participation to government, which might involve establishing independent public participation 'watchdogs' in order to 'monitor and verify that environmental decision making processes are in fact transparent, participatory and accountable' (2006: 193). Edgar (2013) suggests the need for judicial review of public consultation processes as a 'safeguard against tokenism'. Opportunities to review the merits of decisions through independent forums must be available and open to all stakeholders under liberal standing provisions.

A related matter is inclusiveness, and whether the full range of stakeholders are accounted for in participatory processes. A limited selection of participants can lead to perceptions of bias if particular perspectives dominate dialogue, and some stakeholders may be excluded because their interests are not self-evident; both factors can lead to skewed representation. Respectful treatment of all participants is essential to protect trust in governments and proponents. There is also a need to avoid over-reliance on particular individuals and groups as representatives in participatory processes, which creates a risk of 'consultation fatigue', especially where multiple developments are being evaluated (Reed 2008; Simpson, de Loë and Andrey 2015). This is particularly pertinent for rural and Indigenous communities where small population density can result in a limited pool of participants (Howard 2017).

Incorporating place-based knowledge

Place-based impacts are typically poorly evaluated in land use decision making. While some development assessment frameworks require general consideration of social impacts (for example, the NSW *Environmental Planning and Assessment Act 1979*), there is typically little specific guidance provided about how impacts should be evaluated. The quantitative assessment methods (such as cost–benefit analysis) often used to gauge impacts attempt to ascribe a market value to the socio-cultural externalities of development, but communicate little about who actually benefits and who is burdened by development, and how (Cotton 2013; Preston 2015a; Sovacool and Dworkin 2015). Comparisons which 'use money as a proxy for people's quality of life' can neglect people–place relationships (Bronsteen, Buccafusco and Masur 2013: 1689) as well as distort socio-cultural concerns and the distributional effects of development decisions (McManus, Albrecht and Graham 2014; Groves 2015).

The integration of 'vernacular' and place-based knowledge with scientific data can assist in revealing the spectrum of concerns, and encourage the negotiation of solutions that are more likely to be accepted (Bartel 2014; Lin and Lockwood 2014; Simpson et al. 2015). Recent studies have focused on collaborative approaches that facilitate the co-production of place-based knowledge (including societal beliefs and values) and expert science (Simpson et al. 2015; Penny, Williams, Gillespie and Khem 2016). Others have suggested the need to document individual – as well as collective – 'psychoterratic' or environmentally related emotions and mental health states, including solastalgia (as discussed in the Bulga case study in Chapter 3) (McManus et al. 2014). Such approaches recognise the importance of non-scientific understandings of the biophysical environment, providing a conceptual basis and a vocabulary for deliberating place-based concerns in order to 'de-marginalise' non-experts (Bartel 2014: 905; McManus et al. 2014). In turn, this can assist in maintaining place attachment because stakeholders 'perceive themselves as part of the process' (Simpson et al. 2015: 4) and as retaining a sense of control over their future.

A promising method of such integration is social impact assessment (SIA), which is:

> [T]he process of identifying and managing the social issues of project development, and includes the effective engagement of affected communities in participatory processes of identification, assessment and management of social impacts.
>
> (Vanclay, Esteves, Aucamp and Franks 2015: iv)

Social impacts can include positive or negative changes to people's way of life; their culture, community cohesion and political systems; their physical environment; their health and well-being (physical, mental and spiritual); personal and property rights; and their fears and aspirations for their future (Vanclay 2003). SIA mechanisms can assist in moving beyond simplifying quantified assessments of costs and benefits, shifting the focus of evaluation towards whether or not development improves outcomes for individuals and communities (Esteves, Franks and Vanclay 2012). SIA can also encourage developers to work with stakeholders to manage a project's social impacts, and to negotiate 'free, prior and informed consent' for development,[1] formalising a social licence to operate (Esteves et al. 2012; de Rijke 2013).

Esteves et al. note that 'good' SIA practice:

> [I]s participatory; supports affected peoples, proponents, regulatory and support agencies; increases their understanding of how change comes about and increases their capacities to respond to change; and has a broad understanding of social impacts.
>
> (Esteves et al. 2012: 40)

While there is much promise in SIA, its role in development assessment processes has, to date, been relatively minor, and quality control of SIA methods has been poor (Esteves et al. 2012; de Rijke 2013; Preston 2015a). In some cases, assessments have been based upon secondary data, offering only a superficial analysis of the 'spatial, temporal and stakeholder distribution of impacts and benefits' (Esteves et al. 2012: 37). When used by proponents of development, assessments of social impacts may be subject to bias and structured around issues identified by the proponents rather than informed by the community. Thus, the danger is that SIA can become a 'product' for developers to use to legitimise their plans for the purposes of project approval rather than a productive 'process of management' to be engaged in with stakeholders throughout all project phases (Vanclay et al. 2015: iv).

The challenge, therefore, is to find ways to ensure that place-based information is harnessed effectively. Both McManus et al. (2014) and Preston (2015a) recommend an expanded and independent SIA process for development approval, separate from environmental and economic impact assessments and mediated by community empowerment, where affected groups have influence over (and standing in) procedures. A method for doing this may be to extend 'the formal procedures into less formal settings' in which opportunities for community influence and control are increased, and the vulnerable are supported (Preston 2015a: 191). These observations reiterate the importance of meaningful opportunities for public participation to capture and integrate place-based knowledge but also emphasise the importance of ensuring the integrity of the governance system as a whole.

The need for 'meta-governance'

Fundamentally, each of the proposed reforms discussed above point towards the need for a whole-of-system approach to ensure the effectiveness of the governance system. As Jentoft posits, a governance system which is typified by 'inherent inequities and justice disparities among its constituents is likely to experience tensions and conflicts that might lead to a lack of cooperation and resistance to governing interventions' (2013: 45). To increase governability, it is necessary to look at the functioning of the governing system as a whole – not only the design of specific instruments and institutions, but the constitutive principles upon which the system itself is based (Jentoft 2013).

The process of meta-governance – or 'governing how to govern' – can set out overarching 'values, norms and principles' (Kooiman and Jentoft 2009: 819, 823) for the interaction between the various actors and institutional structures within a governance system. It can establish the conditions for good governance by determining the ground rules of the system, reflexively evaluating its effectiveness, and accounting for structural imbalances within the system (Jessop 1998: 43; 2003; Bell and Park 2006). This can improve the ability for governance mechanisms to cope with the difficult choices often faced in land use decisions concerning extractive development. As Kooiman and Jentoff elaborate:

> [I]n the context of hard (substantive) choices . . . such choices can be made less hard when the values, norms and principles guiding governance itself are made coherent and explicit. In other words, when the processes in which substantive issues are formulated and the choices inherent in them are made, are not haphazard ones but guided by an explicit set of meta-governance principles which are deliberated by and made explicit to all concerned, public and private, in an interactive learning context.
>
> (Kooiman and Jentoff 2009: 819)

Environmental justice is a dynamic process 'of working towards environmental justice objectives and against forces which are serving to produce or sustain patterns of injustice' (Walker 2012: 221). In this sense, environmental justice can act as a basis for meta-governance of land use governance systems. As a standard of good governance, the normative principles of environmental justice can inform what is acceptable in terms of the creation and implementation of administrative and policy arrangements (Jentoft 2013).

Because justice (and environmental justice) is a contested notion, negotiating the principles that should guide the governance system as a whole will not be a straightforward task. Nonetheless, if governments make explicit the assumptions which underpin governance choices, and are prepared to scrutinise and debate them through an interactive and transparent deliberative process, power relations can be challenged and opportunities for consensus can emerge (Kooiman and Jentoff 2009: 831). A land use governance system guided by the principles of environmental justice – including equitable outcomes, fair processes, recognition and respectful treatment of all stakeholders, and a commitment to enhance the capabilities integral to individual and collective freedom and functioning – which is supported by robust review procedures, has a better chance of governing with integrity. Ascertaining how the principles of environmental justice might inform the basis of land use governance systems, and how specific justice-based reforms would work within them is thus an area for further theoretical and empirical research.

Conclusion: justice 'as a means and an end'

In the context of growing global energy demand and increasing economic pressures – and with concerns escalating over resource scarcity and environmental sustainability – many new conflicts over extractive development and land use have emerged in communities around the globe. These conflicts are inherently complicated, involving difficult choices between seemingly incommensurate values. They are not amenable to simple resolution, but, as Sovacool and Dworkin have noted, decision makers continue to apply 'routine analyses' to these problems, which do not offer 'suitable answers': 'The enduring questions they [energy development decisions] provoke involve aspects of equity and morality that are seldom explicit in contemporary energy planning and analysis' (2015: 435).

The case studies detailed in this book suggest that the effective governance of extractive resource development requires stronger meta-governance based upon principles of environmental justice. A governance system which is characterised by injustice is likely to result in a refusal to accept governance interventions and is more susceptible to conflict. Across the case studies, it was evident that there was no objective public evaluation of the effectiveness and efficiency of governance systems; there was no articulation of principles to guide the selection and compatibility of particular governance mechanisms (such as whether or not certain administrative decisions should be subject to merits review), nor any performance standards set for their operation. Systemic factors – notably, the power of proponents and government to shape governance systems, and the spatial characteristics of rural places – combined to restrict what issues could be debated and adjudicated. Overall, the failure to ensure the integrity of the governance system contributed to a sense of injustice. This, in turn, led to the criticism of particular land use decisions and, in some cases, fomented social conflict.

Justice cannot be achieved simply through redistribution of risks and harms, or even through isolated reforms to decision making procedures, because this 'leaves intact the process through which . . . problems are created and re-created' (Lake 1996: 171). Evidence from the case studies described in this book and other studies show that governance processes are shaped by the interactions between a variety of actors (including 'individuals, associations, leaders, firms, departments, international bodies, and so on') and structures (including 'culture, law, agreements, material and technical possibilities, and the many other dimensions which constitute the world we live in') (Kooiman and Jentoft 2009: 20). Actors may be enabled or constrained by these structures; they are subject to their influence, but they can also construct and give meaning to them. This is apparent from the case studies in which governance processes played out in particular ways as a result of these overarching system factors, such as access to information and privileging of scientific information. In some instances the factors played out contrary to their intended aims of resolving social conflict over resource development and, in many situations, created injustices where people would not ordinarily expect to be disadvantaged.

Establishing environmental justice as a basis of meta-governance can confront the systemic causes of injustice. In this way, environmental justice is both a 'means and an end' (Gross 2014: 156); concerned with revealing injustice and countering it by informing future governance efforts. By embedding environmental justice as an explicit principle of the overarching land use governance system, a normative framework can be established for the creation and conduct of the institutions that are employed to resolve complex problems. This paves the way for specific reforms that encourage just processes, fair treatment and equitable outcomes. Such reforms would be beneficial to the resolution of conflict over extractive resource development and to many land use and resource allocation conundrums, including native vegetation conservation, renewable energy production, food security and water use.

Against the backdrop of increasing land use conflict, it is critical that the governance of extractive resource development is not only efficient and effective, but also equitable. Governments must lead the charge in this direction. As Gross observes:

> Decision-makers may not have full control over the fairness of decisions, particularly in the case of scarce natural resources. But they *do* have control in the way they treat people, and to a greater or lesser degree, they can set up fair decision-making processes.
>
> (Gross 2014: 153; emphasis in original)

Through an evidence-based analysis of current conflicts, this book has demonstrated that when governance mechanisms fail to meet expectations of justice, claims of *injustice* will arise. Social conflict can ensue particularly if both decision making outcomes *and* processes are perceived as unfair. To minimise the incidence of such conflict and ensure an ethical resolution of extractive development dilemmas, an environmental justice-based approach to land use decision making is required. Principles of environmental justice can result in fairer processes and help produce outcomes that are also seen as fair, reducing the likelihood of conflict. Moreover, by focusing on the capabilities of individuals and communities and what people can actually do and be as a result of extractive development decisions, the interdependence between humans and the environment is prioritised. This is likely to result in decisions which enhance the well-being of individuals and their communities as well as the environment.

Note

1   As McGee notes, while 'free, prior and informed consent' is an established right for Indigenous peoples under international law for extractive development, 'its application to other populations and communities in similar development contexts is an unsettled legal issue' (2009: 572). However, it is emerging in these other contexts as a principle of best practice for extractive development (Voss and Greenspan 2012).

References

ABC Online. (2014, December 31). Gomeroi artefacts to be preserved if mine approved: Shenhua. *ABC News Online*. Retrieved from www.abc.net.au/news/2014-12-31/gomeroi-artefacts-to-be-preserved-if-mine-approved3a-shenhua/5994314

Adger, W. N., Barnett, J., Chapin III, F. S. & Ellemor, H. (2011). This must be the place: underrepresentation of identity and meaning in climate change decision-making. *Global Environmental Politics*, *11*(2), 1–25.

Akerman, P. (2015, July 13). No option but to approve Shenhua mine: Greg Hunt. *The Australian*. Retrieved from www.theaustralian.com.au/business/no-option-but-to-approve-shenhua-mine-greg-hunt/news-story/9fd76e8eaf78a3749aea582a2b736171

Albrecht, G. (2005). 'Solastalgia': a new concept in health and identity. *Philosophy Activism Nature*, *3*(1), 41–55.

Alden, S. (2014, November 5). Due process. *Quirindi Advocate*. Retrieved from www.quirindiadvocate.com.au/due-process/

Alkire, S. (2005). Why the capability approach? *Journal of Human Development*, *6*(1), 115–135.

Anderson, P. (2013). Reasonable accommodation: split estates, conservation easements, and drilling in the Marcellus Shale. *Virginia Environmental Law Journal*, *31*, 136–167.

Andrew, J., & Robottom, I. (2005). Communities' self-determination: whose interests count? Social learning in environmental management: towards a sustainable future. In M. Keen, V. Brown & R. Dyball (Eds.), *Social Learning in Environmental Management: Towards a Sustainable Future* (pp. 63–77). London: Earthscan.

Andrews, E., & McCarthy, J. (2014). Scale, shale, and the state: political ecologies and legal geographies of shale gas development in Pennsylvania. *Journal of Environmental Studies and Sciences*, *4*(1), 7–16.

Aoki, K. (2000). Space invaders: critical geography, the third world in international law and critical race theory. *Villanova Law Review*, *45*, 913–957.

Apple, B. E. (2014). Mapping fracking: an analysis of law, power, and regional distribution in the US. *Harvard Environmental Law Review*, *38*, 217.

Arcioni, E., & Mitchell, G. (2005). Environmental justice in Australia: when the RATS became IRATE. *Environmental Politics*, *14*(3), 363–379.

Argetsinger, B. (2011). Marcellus Shale: bridge to a clean energy future or bridge to nowhere – the environmental, energy and climate policy considerations for shale gas development in New York State. *Pace Environmental Law Review*, *29*, 321–343.

Arnstein, S. R. (1969). A ladder of citizen participation. *Journal of the American Institute of Planners*, *35*(4), 216–224.

Aston, H. (2013, January 11). Miners lobbied O'Farrell to pull the plug on legal centre. *Sydney Morning Herald*. Retrieved from www.smh.com.au/nsw/miners-lobbied-ofarrell-to-pull-the-plug-on-legal-centre-20130110-2cixj.html#ixzz34lxXZaOt

Aston, H. (2016, October 18). The eight-year hitch: controversial Shenhua Watermark coal mine to pass kill date. *Sydney Morning Herald*. Retrieved from www.smh.com.au/federal-politics/political-news/the-8year-hitch-controversial-shenhua-watermark-coal-mine-to-pass-kill-date-20161018-gs4sfp.html

Atkinson, C. M. (2005). *Coal bed methane hazards in New South Wales*. Report prepared for Australian Gas Alliance.

Australian Bureau of Statistics (ABS). (2001). *8414.0 – Australian mining industry, 1998–99*. Retrieved from www.abs.gov.au/ausstats/abs@.nsf/94713ad445ff1425ca2568200019 2af2/93136e734ff62aa2ca2569de00271b10!OpenDocument

Australian Bureau of Statistics (ABS). (2011a). *2001.0 – Census of population and housing: basic community profile, 2011 third release*. Retrieved from www.abs.gov.au/ausstats/abs@.nsf/mf/2001.0

Australian Bureau of Statistics. (ABS). (2011b). *National regional profile: Narrabri (A) (Local Government Area) – Industry*. Retrieved from www.abs.gov.au/AUSSTATS/abs@nrp.nsf/Previousproducts/LGA15750Industry12006-2010?opendocument&tabname=Summary&prodno=LGA15750&issue=2006-2010

Australian Bureau of Statistics (ABS). (2016a). *6291.0.55.003 – Labour force, Australia, Detailed, Quarterly, Feb 2016*. Retrieved from www.abs.gov.au/ausstats/abs@.nsf/mf/6291.0.55.003

Australian Bureau of Statistics (ABS). (2016b). *3218.0 – Regional population growth, Australia, 2014–15*. Retrieved from www.abs.gov.au/ausstats/abs@.nsf/mf/3218.0

Australian Bureau of Statistics (ABS). (2016c). *7503.0 – Value of agricultural commodities produced, Australia, 2014–15*. Retrieved from www.abs.gov.au/ausstats/abs@.nsf/mf/7503.0

Australian Energy Market Operator. (2015). *Gas statement of opportunities for Eastern and South-Eastern Australia*. Retrieved from https://www.aemo.com.au/Gas/National-planning-and-forecasting/~/-/media/D6A46E5B9438431188D39018CED0EB39.ashx

Babbitt, B. (1993). The future environmental agenda for the US. *University of Colorado Law Review, 64*, 513.

Bachrach, P., & Baratz, M. (1970). *Power and Poverty: Theory and Practice*. London: Oxford University Press.

Ballet, J., Koffi, J. & Pelenc, J. (2013). Environment, justice and the capability approach. *Ecological Economics, 85*, 28–34.

Bartel, R. (2014). Vernacular knowledge and environmental law: cause and cure for regulatory failure. *Local Environment, 19*(8), 891–914.

Bartel, R., & Graham, N. (2016). Property and place attachment: a legal geographical analysis of biodiversity law reform in New South Wales. *Geographical Research, 54*(3), 267–284.

Bartel, R., Graham, N., Jackson, S., Prior, J. H., Robinson, D., Sherval, M. & Williams, S. (2013). Legal geography: an Australian perspective. *Geographical Research, 51*(4), 339–353.

Bartel, R., McFarland, P. & Hearfield, C. (2014). Taking a de-binarised envirosocial approach to reconciling the environment vs economy debate: lessons from climate change litigation for planning in NSW, Australia. *Town Planning Review, 85*(1), 67–96.

Battersby, L. (2013, May 13). Hazzard joins challenge to mine expansion. *Sydney Morning Herald*. Retrieved from www.smh.com.au/business/hazzard-joins-challenge-to-mine-expansion-20130512-2jg6f.html

Bell, D. (2004). Environmental justice and Rawls' difference principle. *Environmental Ethics, 26*(3), 287–306.

Bell, K. (2014). *Achieving environmental justice. A cross-national Analysis.* Bristol: Policy Press.

Bell, S., & Park, A. (2006). The problematic metagovernance of networks: water reform in New South Wales. *Journal of Public Policy, 26*(01), 63–83.

Bennett, L., & Layard, A. (2015). Legal geography: becoming spatial detectives. *Geography Compass, 9*(7), 406–422.

Benson, M. H. (2014). The rules of engagement: the spatiality of judicial review. In I. Braverman, D. Delaney, A. Kedar & N. Blomley (Eds.), *Expanding the Spaces of Law* (pp. 215–238). Palo Alto, CA: Stanford University Press.

Bernal, A. (2011). Power, powerlessness and petroleum: Indigenous environmental claims and the limits of transnational law. *New Political Science, 33*(2), 143–167.

BHP Billiton. (2010, March 24). BHP Billiton Statement: blockade on exploration activities lifted. *BHP Billiton.* Retrieved from www.abc.net.au/reslib/201003/r537037_3095841.pdf

BHP Billiton. (2014). *Caroona Coal Project: project description and preliminary environmental assessment.* Retrieved from https://majorprojects.affinitylive.com/public/354adfc3c d5e3d56dc2558950cf3083a/C%20aroona%20Coal%20Project%20%20%20Preliminary%20Environmental%20Assessment.pdf

Bickerstaff, K., & Agyeman, J. (2009). Assembling justice spaces: the scalar politics of environmental justice in North-East England. *Antipode, 41*(4), 781–806.

Biernacki, P., & Waldorf, D. (1981). Snowball sampling: problems and techniques of chain referral sampling. *Sociological methods & research, 10*(2), 141–163.

Bies, R. (2001). Interactional (in)justice: the sacred and the profane. In J. Greenberg & R. Cropanzano (Eds.), *Advances in Organizational Justice.* Mahwah, NJ: Lawrence Erlbaum.

Bita, N. (2011, June 27). Chinese mine giant snaps up 43 NSW farms. *The Australian.* Retrieved from www.theaustralian.com.au/news/nation/chinese-mine-giant-snaps-up-43-nsw-farms/story-e6frg6nf-1226082387428

Blomley, N., Delaney, D. & Ford, R. T. (2001). Preface: Where is law? In N. Blomley, D. Delaney & R. T. Ford (Eds.), *The Legal Geographies Reader* (pp. xiii–xxii). Oxford: Blackwell Publishers.

Bohman, J. (2007) Beyond distributive justice and struggles for recognition: freedom, democracy, and critical theory. *European Journal of Political Theory, 6*(3), 267–276.

Bourdieu, P. (1987). The force of law: toward a sociology of the juridical field. *Hastings Law Journal, 38*, 805–1297.

Brasier, K. J., Filteau, M. R., McLaughlin, D. K., Jacquet, J., Stedman, R. C., Kelsey, T. W. & Goetz, S. J. (2011). Residents' perceptions of community and environmental impacts from development of natural gas in the Marcellus Shale: a comparison of Pennsylvania and New York cases. *Journal of Rural Social Sciences, 26*(1), 32.

Braun, V., & Clarke, V. (2006). Using thematic analysis in psychology. *Qualitative Research in Psychology, 3*(2), 77–101.

Brehm, H. N., & Pellow, D. N. (2013). Environmental justice: pollution, poverty, and marginalized communities. In P. G. Harris (Ed.), *Routledge Handbook of Global Environmental Politics* (pp. 308–320). Abingdon: Routledge.

Bronsteen, J., Buccafusco, C. & Masur, J. S. (2013). Well-being analysis vs. cost-benefit analysis. *Duke Law Journal, 62*, 1603–1689.

Browning, J., & Kaplan, A. (2011). *Deep drilling, deep pockets – in congress & Pennsylvania.* Washington, DC: Common Cause.

Bryner, G. (2002). Assessing claims of environmental justice: conceptual frameworks. In K. Mutz, G. Bryner & D. Kennedy (Eds.), *Justice and Natural Resources: Concepts, Strategies, and Applications* (pp. 31–56). Washington, DC: Island Publishers.

Bulga Milbrodale Progress Association (BMPA). (2016). *Bulga Milbrodale Progress Association: Statement*. Retrieved from https://d3n8a8pro7vhmx.cloudfront.net/lockthegate/pages/3401/attachments/original/1463689652/BMPA_Statement.pdf?1463689652

Bullard, R. (1990). *Dumping in Dixie: Race, Class, and Environmental Quality*. Boulder, CO: Westview Press.

Bullard, R. (2002). Confronting environmental racism in the twenty-first century. *Global Dialogue, 4*(1), 34–48.

Bureau of Resources and Energy Economics (BREE). (2012). *Resources and energy statistics*. Canberra, Australia: Commonwealth of Australia.

Campbell, R. (2014). *Seeing through the dust: coal in the Hunter Valley economy*. Retrieved from www.tai.org.au/content/seeing-through-dust-coal-hunter-valley-economy

Caroona Coal Action Group (CCAG). (2015a). *Shenhua Watermark Open Cut Coal Mine: summary of recommendations from PAC 1 and the final PAC determination*. Retrieved from http://ccag.org.au/wp-content/uploads/2015/03/Summary-of-PAC-recs-and-determination.pdf

Caroona Coal Action Group (CCAG). (2015b). *PAC process fails to address community concerns*. Retrieved from http://ccag.org.au/pac-process-fails-to-address-community-concerns/

Caroona Coal Action Group (CCAG). (n.d.). *About us*. Retrieved from http://ccag.org.au/about-us/

Carrington, K., & Pereira, M. (2011). Assessing the social impacts of the resources boom on rural communities. *Rural Society, 21*(1), 2–20.

Cartoscope. (n.d.). *Hunter Valley Coal Heritage*. Retrieved from Geological sites of NSW website: www.geomaps.com.au/scripts/huntervalleycoal.php

Chambers, M. (2015, February 13). Santos chief defends CSG foray in NSW despite $808m impairment. *The Australian*. Retrieved from www.theaustralian.com.au/business/mining-energy/santos-chief-defends-csg-foray-in-nsw-despite-808m-impairment/news-story/4aecf44f82b871072afa17b6859d9570

Chen, C., & Randall, R. (2013). The economic contest between coal seam gas mining and agriculture on prime farmland: it may be closer than we thought. *Journal of Economic and Social Policy, 15*(3), Article 5. Retrieved from www.epubs.scu.edu.au/jesp/vol15/iss3/5/

Chillingworth, B. (2014, April 30). Protesters getting stitched up in Narrabri. *Northern Daily Leader*. Retrieved from www.northerndailyleader.com.au/story/2248488/protesters-getting-stitched-up-in-narrabri/

Chung, C. K. L., & Xu, J. (2015). Scale as both material and discursive: a view through China's rescaling of urban planning system for environmental governance. *Environment and Planning C: Government and Policy, 34*(8), 1404–1424.

Clark, C. P. (2016). The politics of public interest environmental litigation: Lawfare in Australia. *Australian Environment Review, 31*(7), 258–262.

Clayton, S., & Opotow, S. (2003). Justice and identity: changing perspectives on what is fair. *Personality and Social Psychology Review, 7*(4), 298–310.

Clayton, S., Koehn, A. & Grover, E. (2013). Making sense of the senseless: identity, justice, and the framing of environmental crises. *Social Justice Research, 26*(3), 301–319.

Cleary, P. (2012). *Mine-field: The Dark Side of Australia's Resources Rush*. Carlton, Victoria: Black.

Cleary, P. (2015, August 4). Farmers against Whitehaven and Shenhua in battle of the plains. *The Australian*. Retrieved from www.theaustralian.com.au/news/inquirer/farmers-against-whitehaven-and-shenhua-in-battle-of-the-plains/news-story/f9d4d785bcab3f929b8d16cfcdf5bed9

Clifford, C. (2011, July 5). Protester occupies Pilliga CSG well. *ABC News Online*. Retrieved from www.abc.net.au/news/2011-07-05/protester-occupies-pilliga-csg-well/2783282

Clifford, C. (2014, February 5). Santos machinery halted by protestor sit-in. *ABC News Online*. Retrieved from www.abc.net.au/news/2014-02-05/santos-machinery-halted-by-protestor-sit-in/5241018

Climate and Health Alliance. (2015). *Coal and health in the Hunter: lessons from one valley for the world*. Retrieved from http://caha.org.au/wp-content/uploads/2010/01/Climate-and-Health-Alliance_Report_Layout_PRINTv2.pdf

Clough, E., & Bell, D. (2016). Just fracking: a distributive environmental justice analysis of unconventional gas development in Pennsylvania, USA. *Environmental Research Letters, 11*(2), article id. 025001.

Coal & Allied. (2013, April 22). *Media release: Coal & Allied appeals Warkworth extension rejection*. Retrieved from www.riotinto.com/documents/MediaReleases-coal-australia/130422_MR_Warkworth_extension_Supreme_Court_appeal2.pdf

Colagiuri, R., Cochrane, J. & Girgis, S. (2012). *Health and social harms of coal mining in local communities: spotlight on the Hunter region*. Retrieved from https://sydney.edu.au/medicine/research/units/boden/PDF_Mining_Report_FINAL_October_2012.pdf

Cole, L., & Foster, S. (2001). *From the Ground Up: Environmental Racism and the Rise of the Environmental Justice Movement*. New York: New York University Press.

Colvin, R., Witt, G. B. & Lacey, J. (2015). Strange bedfellows or an aligning of values? Exploration of stakeholder values in an alliance of concerned citizens against coal seam gas mining. *Land Use Policy, 42*, 392–399.

Condon, M. (2010, March 11). Minerals Council says court decision could severely curtail exploration and jobs. *ABC Rural Online*. Retrieved from www.abc.net.au/site-archive/rural/nsw/content/2010/03/s2843212.htm

Coni-Zimmer, M., Flohr, A. & Jacobs, A. (2016). Claims for local justice in natural resource conflicts. In M. Pichler, C. Staritz, K. Küblböck, C. Plank, W. Raza & F. R. Peyré (Eds.), *Fairness and Justice in Natural Resource Politics* (pp. 90–108). London: Routledge.

Connor, L., Albrecht, G., Higginbotham, N., Freeman, S. & Smith, W. (2004). Environmental change and human health in Upper Hunter communities of New South Wales, Australia. *EcoHealth, 1*(2), SU47–SU58.

Conrad, E., Cassar, L. F., Christie, M. & Fazey, I. (2011). Hearing but not listening? A participatory assessment of public participation in planning. *Environment and Planning C: Government and Policy, 29*(5), 761–782.

Cook, J. J. (2015). Who's pulling the fracking strings? Power, collaboration and Colorado fracking policy. *Environmental Policy and Governance, 25*(6), 373–385.

Corbett, T. (2011, March 8). Governor Tom Corbett 2011–12 Budget Address. *PR Newswire*. Retrieved from www.prnewswire.com/news-releases/governor-tom-corbett-2011-12-budget-address-117586608.html

Corbett, T. (2012). *Executive Order, Commonwealth of Pennsylvania: Permit Decision Guarantee for the Department of Environmental Protection*. Retrieved from www.portal.state.pa.us/portal/server.pt?open=512&objID=708&PageID=224602&mode=2&contentid=http://pubcontent.state.pa.us/publishedcontent/publish/cop_general_government_operations/oa/oa_portal/omd/p_and_p/executive_orders/2010_201

Cosgrove, B., LaFave, D. R., Donihue, M. R. & Dissanayake, S. (2015). The economic impact of shale gas development: a natural experiment along the New York /Pennsylvania border. *Agricultural and Resource Economics Review, 44*(2), 20–39.

Cottle, D. (2013). Land, life and labour in the sacrifice zone: the socio-economic dynamics of open-cut coal mining in the Upper Hunter Valley, New South Wales. *Rural Society 22*(3), 208–216.

Cottle, D., & Keys, A. (2014). Open-cut coal mining in Australia's Hunter Valley: sustainability and the industry's economic, ecological and social implications. *International Journal of Rural Law and Policy Special edition, 1*, 1–7.

Cotton, M. (2013). Shale gas–community relations: NIMBY or not? Integrating social factors into shale gas community engagements. *Natural Gas & Electricity, 29*(9), 8–12.

Cotton, M. (2016). Fair fracking? Ethics and environmental justice in United Kingdom shale gas policy and planning. *Local Environment, 22*(2), 185–202.

CSIRO. (2015). *What is unconventional gas?* Retrieved from www.csiro.au/en/Research/Energy/Hydraulic-fracturing/What-is-unconventional-gas

Cubby, B. (2013, December 22). Legal aid cuts a blow for anti-gas groups. *Sydney Morning Herald*. Retrieved from www.smh.com.au/national/legal-aid-cuts-a-blow-for-antigas-groups-20121221-2brgt.html

Cubby, B., & Nicholls, S. (2011, November 15). Win for gas blockade as Santos pulls back. *Sydney Morning Herald*. Retrieved from www.smh.com.au/environment/win-for-gas-blockade-as-santos-pulls-back-20111114-1nfk7.html

Cubby, B., & Nicholls, S. (2012, January 31). Food zones for farmers in danger of coming a cropper. *Sydney Morning Herald*. Retrieved from www.smh.com.au/environment/conservation/food-zones-for-farmers-in-danger-of-coming-a-cropper-20120130-1qppp.html

Cubby, B., & Rigney, S. (2013, April 16). Tiny Bulga wins day against mining Goliath, *Sydney Morning Herald*. Retrieved from www.smh.com.au/environment/conservation/tiny-bulga-wins-day-against-mining-goliath-20130415-2hw5n.html#ixzz489DdKkOe

Cunningham, P. M. (1827). *Two Years in New South Wales; a Series of Letters, Comprising Sketches of the Actual State of Society in that Colony; of its Peculiar Advantages to Emigrants; of its Topography, Natural History, etc.* (No. 31). London: H. Colburn.

Cusick, M. (2015a, January 22). Anti-drilling group settles surveillance litigation with state. *StateImpact*. Retrieved from https://stateimpact.npr.org/pennsylvania/2015/01/page/2/

Cusick, M. (2015b, February 2) In fracking hot spots, police and gas industry share intelligence on activists. *StateImpact*. Retrieved from https://stateimpact.npr.org/pennsylvania/2015/02/02/in-fracking-hot-spots-police-and-gas-industry-share-intelligence-on-activists/

Dahl, R. (1957). The concept of power. *Behavioural Science, 2*(3), 201–215.

Daly, E., & May, J. R. (2015). Robinson Township v. Pennsylvania: a model for environmental constitutionalism. *Widener Law Review, 21*, 151–170.

Davies, A. (2016a, February 20). Dust policy for mines was battleground between EPA and the industry. *Sydney Morning Herald*. Retrieved from www.smh.com.au/nsw/dust-policy-for-mines-was-battleground-between-epa-and-the-industry-20160217-gmwrxw#ixzz48Cg05Cth

Davies, A. (2016b, February 21). Mining industry lobbied to change coal mine acquisition policy, say activists. *Sydney Morning Herald*. Retrieved from www.smh.com.au/nsw/mining-industry-lobbied-to-change-coal-mine-acquisition-policy-say-activists-20160218-gmxgie.html#ixzz48CgHx1C9

Davoudi, S., & Brooks, E. (2014). When does unequal become unfair? Judging claims of environmental injustice. *Environment and Planning A, 46*(11), 2686–2702.

de Rijke, K. (2013). Coal seam gas and social impact assessment: an anthropological contribution to current debates and practices. *Journal of Economic and Social Policy, 15*(3), Article 3. Retrieved from www.epubs.scu.edu.au/jesp/vol15/iss3/.

Delaney, D. (2010). *The Spatial, the Legal and the Pragmatics of World Making: Nomospheric Investigations*. New York: Routledge.

Denning, A. (1955). *The Road to Justice*. London: Stevens and Sons.

Department of Environment and Climate Change. (2008). *Review of State Conservation Areas: report of the first five-year review of State Conservation Areas under the National Parks and Wildlife Act 1974*. Retrieved from www.environment.nsw.gov.au/resources/parks/08516SCAreview.pdf

Department of Environmental Protection. (n.d.). *Office of Environmental Justice*. Retrieved from www.dep.pa.gov/PublicParticipation/OfficeofEnvironmentalJustice/Pages/default.aspx

Department of Planning. (2008). *Project approval: Narrabri Coal Seam Gas Utilisation Project (Wilga Park Power Station)*. Retrieved from https://majorprojects.affinitylive.com/public/7c4704dc3bd9c71a72cf44e10cb75d/Project%20Approval.pdf

Department of Planning and Environment. (2014a). *Secretary's Environmental Assessment Report: Watermark Coal Project (State Significant Development Assessment SSD 4975)*. Retrieved from https://majorprojects.affinitylive.com/public/5a2cf0939ad5112e7a3a99d4f4f0df82/Watermark%20Coal%20Project%20-%20Assessment%20Report.pdf

Department of Planning and Environment. (2014b). *Addendum: State Significant Development Assessment – Watermark Coal Project*. Retrieved from https://majorprojects.affinitylive.com/public/8576ea30f663c7f711919fd872f54644/Watermark%20Assessment%20Report_Addendum.pdf

Department of Planning and Environment. (2015). *Draft change to mining policy – frequently asked questions*. Retrieved from https://majorprojects.affinitylive.com/public/84d84b147f76928b6c64356fb398ec03/Frequently%20Asked%20Questions.pdf

Department of Planning and Infrastructure. (2012). *Strategic Regional Land Use Plan – New England North West*. Retrieved from https://www.nsw.gov.au/sites/default/files/initiatives/newengnw-strategic-plan_sd_v01.pdf

Department of Planning and Infrastructure. (2014). *Fact Sheet: frequently asked questions: biophysical strategic agricultural land mapping across NSW*. Retrieved from www.planning.nsw.gov.au/Policy-and-Legislation/Mining-and-Resources/~/media/C53910CA79E14DE6A2B69134D2429A51.ashx

Department of the Environment. (2013). *Notification of referral decision: Energy NSW Coal Seam Gas Exploration and Appraisal Program, Gunnedah Basin, NSW (EPBC 2013/6918)*. Retrieved from www.environment.gov.au/epbc/notices/assessments/2013/6918/2013-6918-referral-decision.pdf

Department of the Environment. (2015a). *Approval decision: Watermark Coal Project NSW (EPBC 2011/6201)*. Retrieved from www.environment.gov.au/epbc/notices/assessments/2011/6201/2011-6201-approval-decision.pdf

Department of the Environment. (2015b). *Statement of reasons for a decision to approve an action under the Environment Protection and Biodiversity Conservation Act 1999 (Cth)*. Retrieved from www.environment.gov.au/epbc/notices/assessments/2011/6201/2011-6201-statement-of-reasons-s130.pdf

Dernbach, J. (2015). The potential meanings of constitutional public trust. *Environmental Law*, *45*(2), 463–518

Dernbach, J. C., May, J. R. & Kristl, K. T. (2015). Robinson Township v. Commonwealth of Pennsylvania: examination and implications. *Rutgers Law Review*, *67*(1169), 1172–1196.

Devine-Wright, P. (2011a). Place attachment and public acceptance of renewable energy: a tidal energy case study. *Journal of Environmental Psychology*, *31*, 336–343.

Devine-Wright, P. (Ed.). (2011b). *Renewable Energy and the Public: From NIMBY to Participation*. New York: Earthscan.

Devine-Wright, P. (2012). Explaining 'NIMBY' objections to a power line: the role of personal, place attachment and project-related factors. *Environment and Behavior, 45*(6), 761–781.

Dietz, S., & Atkinson, G. (2005). Public perceptions of equity in environmental policy: traffic emissions policy in an English urban area. *Local Environment, 10*(4), 445–459.

Duus, S. (2013). Coal contestations: learning from a long, broad view. *Rural Society, 22*(2), 96.

Dwyer, G. J. (2014). The operation and interplay of polycentricity, procedural fairness and expert opinion evidence in the context of environmental dispute resolution. *Australasian Journal of Natural Resources Law and Policy, 17*(2), 217–261.

Earth Systems. (2013). *Independent Review of the Environmental Impact Statement for the Watermark Coal Project*. Retrieved from https://majorprojects.affinitylive.com/public/2 dd7499182476fda18313154a90c44df/02.%20Watermark%20Coal%20Project%20%20 Carrona%20Coal%20Action%20Group%20-%20Attachment%201.pdf

Eastern Star Gas. (2010). *Narrabri Coal Seam Gas Project – Gas Field Development, Preliminary Environmental Assessment prepared by AECOM Australia Pty Ltd*. Retrieved from https://majorprojects.affinitylive.com/public/e41d04ed8c420d4841248d7e3309a4ec/ 60153583_PEA_07Sep10.pdf

Eaton, E., & Kinchy, A. (2016). Quiet voices in the fracking debate: ambivalence, non-mobilization, and individual action in two extractive communities (Saskatchewan and Pennsylvania). *Energy Research & Social Science, 20*(October), 22–30.

EcoLogical Australia (2011). *Proposed framework for assessing the cumulative risk of mining on natural resource assets in the Namoi catchment*. Retrieved from http://education.nwlls. com/client/multimedia/namoi__risk_assessment_final_v5_14sept11.pdf

Edgar, A. (2013). Judicial review of public consultation processes: a safeguard against tokenism? *Public Law Review, 24*, 209–224.

Eisenberg, A. M. (2015). Beyond science and hysteria: reality and perceptions of environmental justice concerns surrounding Marcellus and Utica shale gas development. *University of Pittsburgh Law Review, 77*(2), 183–234.

Environment Protection Authority (EPA). (2012, July 6). *Media release: Eastern Star Gas fined for pollution in the Pilliga*. Retrieved from www.epa.nsw.gov.au/epamedia/epamedia12070601.htm

Environment Protection Authority (EPA). (2014, February 18). *Media release: Santos fined $1,500 for water pollution*. Retrieved from www.epa.nsw.gov.au/epamedia/EPAMedia14021802.htm

Environmental Defender's Office (EDO). (2011). *Ticking the box: flaws in the environmental assessment of coal seam gas exploration activities*. Retrieved from www.edonsw.org.au/ ticking_the_box_flaws_in_the_environmental_assessment_of_coal_seam_gas_explo ration_activities

Environmental Defender's Office (EDO). (2013). *NSW Environmental Defender's Office Submission on amendments to the Mining SEPP – State Environmental Planning Policy (Mining, Petroleum Production and Extractive Industries) Amendment (Resource Significance) 2013*. Retrieved from http://d3n8a8pro7vhmx.cloudfront.net/edonsw/pages/335/ attachments/original/1380680264/130809MiningSEPPamendments.pdf?1380680264

Environmental Defender's Office (EDO). (2014). *Submission on the Draft NSW Biodiversity Offsets Policy for Major Projects May 2014*. Retrieved from http://d3n8a8pro7vhmx.cloud-front.net/edonsw/pages/1455/attachments/original/1400219519/140516_NSW_Biodi-versity_Offsets_Policy_for_Major_projects_-EDO_NSW_Submission.pdf?1400219519

Esteves, A. M., Franks, D. & Vanclay, F. (2012). Social impact assessment: the state of the art. *Impact Assessment and Project Appraisal, 30*(1), 34–42.

Evans, G. R. (2008). Transformation from 'Carbon Valley' to a 'Post-Carbon Society' in a climate change hot spot: the coalfields of the Hunter Valley, New South Wales, Australia. *Ecology and Society*, 13(1), 39. Retrieved from www.ecologyandsociety.org/vol13/iss1/art39/

Ey, M., & Sherval, M. (2016). Exploring the minescape: engaging with the complexity of the extractive sector. *Area*, 48(2), 176–182.

Fairclough, N. (2013). *Critical Discourse Analysis: The Critical Study of Language*. New York: Routledge.

Farr, B. (2012, August 2). Water study in but opinions divided. *The Land*. Retrieved from www.theland.com.au/story/3604161/water-study-in-but-opinions-divided/?cs=4963

Fershee, J. (2014). Facts, fiction, and perception in hydraulic fracturing: illuminating Act 13 and *Robinson Township v. Commonwealth of Pennsylvania*. *West Virginia Law Review*, 116, 819–863.

Figueroa, R. (2003). Bivalent environmental justice and the culture of poverty. *Rutgers University Journal of Law and Urban Policy*, 1(1), 27–43.

Finley, B. (2015, January 31). Wolf restores fracking ban in state parkland. *The Philadelphia Inquirer*. Retrieved from www.philly.com/philly/hp/news_update/20150130_Wolf_restores_fracking_ban_in_state_parkland.html

Fisher, K. (2015). Communities in the dark: the use of state sunshine laws to shed light on the fracking industry. *Boston College Environmental Affairs Law Review*, 42, 99–131.

Flyvbjerg, B. (2006). Five misunderstandings about case-study research. *Qualitative Inquiry*, 12(2), 219–245.

Foley, M. (2013, October 3). No gate in Gateway: Simson. *The Land*. Retrieved from www.theland.com.au/news/agriculture/general/news/no-gate-in-gateway-simson/2673757.aspx

Foley, M. (2014a, April 17). BHP's floodplain backflip. *The Land*. Retrieved from www.theland.com.au/story/3576583/bhps-floodplain-backflip/?cs=4951#

Foley, M. (2014b, April 7). Pilliga protests roll on. *The Land*. Retrieved from www.theland.com.au/story/3577066/pilliga-protests-roll-on/

Foley, M. (2015a, January 31). Government backs Shenhua against backlash. *The Land*. Retrieved from www.theland.com.au/story/3370539/government-backs-shenhua-against-backlash/

Foley, M. (2015b, April 13). Shenhua mine approval fail. *The Land*. Retrieved from www.theland.com.au/story/3367849/shenhua-mine-approval-fail/

Foley, M. (2016, August 11). BHP buggers off from Caroona coal project. *The Land*. Retrieved from www.theland.com.au/story/4091294/bhp-buggers-off-from-caroona-coal-all-eyes-on-shenhua/?cs=4941

Foote, N. L. (2015). Not in my backyard: unconventional gas development and local land use in Pennsylvania and Alberta, Canada. *Penn State Journal of Law and International Affairs*, 3(2), 235–263.

Foreman Jr., C. (1998). *The Promise and Peril of Environmental Justice*. Washington, DC: Brookings Institution Press.

Fraser, N. (1998). Social justice in the age of identity politics: redistribution, recognition, and participation. In G. Peterson (Ed.), *The Tanner Lectures on Human Values 19* (pp. 1–67). Salt Lake City: University of Utah Press.

Fraser, N. (2000). Rethinking recognition. *New Left Review*, 3, 107–120.

Friends of the Pilliga. (2011a, June 1). *Media release: Eastern Star Gas must come clean on full plans*. Retrieved from Stop Pilliga Coal Seam Gas website: www.stoppilligacoalseamgas.com/eastern-star-gas-must-come-clean/

Friends of the Pilliga (2011b, 21 July). *Media release: 'Claytons' moratorium won't fly in the bush*. Retrieved from Stop Pilliga Coal Seam Gas website: www.stoppilligacoalseamgas. com/claytons-moratorium-wont-fly-in-the-bush/

Friends of the Pilliga. (2012a, July 6). *Media release: EPA fines gas companies for polluting creek with CSG wastewater*. Retrieved from Stop Pilliga Coal Seam Gas website: www.stoppil-ligacoalseamgas.com/epa-fines-gas-companies-for-polluting-creek-with-csg-wastewater/

Friends of the Pilliga. (2012b, June 18). *Media release: What are they hiding? Santos objects to release of 34 documents under FOI*. Retrieved from Stop Pilliga Coal Seam Gas web-site: www.stoppilligacoalseamgas.com/what-are-they-hiding-santos-objects-to-release-of-34-documents-under-foi/

Friends of the Pilliga. (2012c, October 19). *Media release: Show us the maps: Santos keeps community in the dark*. Retrieved from Stop Pilliga Coal Seam Gas website: www.stoppil-ligacoalseamgas.com/show-us-the-maps-santos-keeps-community-in-the-dark/

Friends of the Pilliga. (n.d.). *Coal seam gas threat*. Retrieved from www.stoppilligacoal-seamgas.com/pages/

Fry, M., Briggle, A. & Kincaid, J. (2015). Fracking and environmental (in)justice in a Texas city. *Ecological Economics, 117*, 97–107.

Fuller, K. (2014, December 15). Fears artefacts could be destroyed if Shenhua mine plan is approved. *ABC New England North West*. Retrieved from www.abc.net.au/local/sto-ries/2014/12/12/4147339.htm

Gaventa, J. (1980). *Power and Powerlessness: Quiescence and Rebellion in an Appalachian Valley*. Urbana: University of Illinois Press.

Getches, D. H., & Pellow, D. N. (2002). Beyond 'traditional' environmental justice. In K. Mutz, G. Bryner & D. Kennedy (Eds.), *Justice and Natural Resources: Concepts, Strate-gies, and Applications* (pp. 3–30). Washington, DC: Island Publishers.

Godden, L. C., Langton, M., Mazel, O. & Tehan, M. (2008). Indigenous and local peoples and resource development: international comparisons of law, policy and practice. *Policy and Practice. Journal of Energy & Natural Resources Law, 26*(1), 1–173.

Godfrey, M. (2016, February 8). CSG licence overhaul planned to prevent gas shortages and drive down the price. *The Daily Telegraph*. Retrieved from www.dailytelegraph.com.au/news/nsw/csg-licence-overhaul-planned-to-prevent-gas-shortages-and-drive-down-the-price/news-story/b06dba3a2551e65a5c2f9bab77ae63bc

González, S. (2006). Scalar narratives in Bilbao: a cultural politics of scales approach to the study of urban policy. *International Journal of Urban and Regional Research, 30*(4), 836–857.

Graham, N. (2010). *Lawscape: Property, Environment, Law*. New York: Routledge.

Green D., Petrovic J., Moss P. & Burrell M. (2011). *Water resources and management over-view: Namoi catchment*. Retrieved from NSW Office of Water website: www.water.nsw.gov.au/__data/assets/pdf_file/0003/549300/catchment_overview_namoi.pdf

Griewald, Y., & Rauschmayer, F. (2014). Exploring an environmental conflict from a capa-bility perspective. *Ecological Economics, 100*, 30–39.

Gross, C. (2007). Community perspectives of wind energy in Australia: the application of a justice and community fairness framework to increase social acceptance. *Energy Policy, 35*(5), 2727–2736.

Gross, C. (2008). A measure of fairness: an investigative framework to explore perceptions of fairness and justice in real-life social conflict. *Human Ecology Review, 15*(2), 130.

Gross, C. (2014). *Fairness and Justice in Environmental Decision Making: Water under the Bridge*. New York: Earthscan.

Groves, C. (2015). The bomb in my backyard, the serpent in my house: environmental justice, risk, and the colonisation of attachment. *Environmental Politics, 24*(6), 853–873.

Hagen-Frederiksen, C. C. (2016). Beyond fracking: How Robinson Township alters Pennsylvania municipal zoning rights. *Journal of Law and Commerce, 34*, 375–397.

Hannam, P. (2014a, December 15). Rio Tinto linked to collusion on Warkworth coalmine. *Sydney Morning Herald*. Retrieved from www.smh.com.au/environment/rio-tinto-linked-to-collusion-on-warkworth-coalmine-20141212-1267rh.html

Hannam, P. (2014b, August 21). Rio Tinto 'workshopped' Warkworth coal plan with Planning Department, Greens say. *Sydney Morning Herald*. Retrieved from www.smh.com.au/environment/rio-tinto-workshopped-warkworth-coal-plan-with-planning-department-greens-say-20140820–106bdx.html

Hannam, P. (2014c, July 11). BHP's Caroona coal mine fails Gateway tests. *Sydney Morning Herald*. Retrieved from www.smh.com.au/environment/water-issues/bhps-caroona-coal-mine-fails-gateway-tests-20140711-zt47v.html

Hannam, P. (2015a, November 22). Plea for Premier Mike Baird to reverse Rio Tinto's 'original sin' on Warkworth mine. *Sydney Morning Herald*. Retrieved from www.smh.com.au/environment/plea-for-premier-mike-baird-to-reverse-rio-tintos-original-sin-on-warkworth-mine-20151111-gkwxta.html#ixzz48KB5snw3

Hannam, P. (2015b, July 8). Giant Shenhua Watermark coal mine wins federal approval from Environment Minister Greg Hunt. *Sydney Morning Herald*. Retrieved from www.smh.com.au/environment/giant-shenhua-watermark-coal-mine-wins-federal-approval-from-environment-minister-greg-hunt-20150708-gi7j65.html

Hannam, P. (2016a, August 11). Liverpool Plains: Baird government to pay BHP Billiton $220 million for licence. *Sydney Morning Herald*. Retrieved from www.smh.com.au/environment/liverpool-plains-baird-government-to-pay-bhp-billiton-220-million-for-licence-20160811-gqqhkv.html

Hannam, P. (2016b, July 31). Documents reveal 'close and cosy' CSG relationship between Santos and government. *Sydney Morning Herald*. Retrieved from www.smh.com.au/environment/documents-reveal-close-and-cosy-csg-relationship-between-santos-and-government-20160728-gqfhzg.html

Hardy, B. (2012, October 23). Namoi Catchment Water Study 'gathering dust'. *Northern Daily Leader*. Retrieved from www.northerndailyleader.com.au/story/416581/namoi-catchment-water-study-gathering-dust/

Hardy, K., & Kelsey, T. W. (2015). Local income related to Marcellus Shale activity in Pennsylvania. *Community Development, 46*(4), 329–340.

Hartcher, The Hon Chris (Minister for Resources and Energy). (2011, July 21). *Media release: NSW Govt has listened and acted: tough new conditions for coal & coal seam gas*. Retrieved from www.resourcesandenergy.nsw.gov.au/__data/assets/pdf_file/0005/408623/Tough-new-conditions-for-coal-exploration.pdf

Harvey, D. (1996). *Justice, Nature and the Geography of Distance*. Oxford: Blackwell.

Hasham, N. (2014, April 12). Up to their necks in it, farmers lead coal seam gas protests by example. *Newcastle Herald*. Retrieved from www.theherald.com.au/story/2214618/up-to-their-necks-in-it-farmers-lead-coal-seam-gas-protests-by-example/

Hasham, N. (2016, March 3). No evidence of 'vigilante' green groups as government crackdown falls silent. *Sydney Morning Herald*. Retrieved from www.smh.com.au/federal-politics/political-news/no-evidence-of-vigilante-green-groups-as-government-crackdown-falls-silent-20160303-gn9e7c.html

Hazzard, The Hon. Brad MP (Minister for Planning and Infrastructure). (2011, May 21). *Media release: NSW Government adopts rigorous strategic approach to regional land use planning*. Retrieved from www.dpi.nsw.gov.au/__data/assets/pdf_file/0007/390976/NSW-Government-adopts-rigorous-strategic-approach-to-regional-land-use-planning.pdf

Hepburn, S. (2013). The importance of the federal coal seam gas water trigger. *Australian Environment Review, 28*(6), 612–616.

Heynen, N. C., Kaika, M. & Swyngedouw, E. (2006). *In the Nature of Cities: Urban Political Ecology and the Politics of Urban Metabolism* (Vol. 3). New York: Taylor & Francis.

Higginbotham, N., Freeman, S., Connor, L. & Albrecht, G. (2010). Environmental injustice and air pollution in coal affected communities, Hunter Valley, Australia. *Health & Place, 16*(2), 259–266.

Higgins, J. (2013). Fracking the Marcellus Shale. *George Washington Journal of Energy and Environmental Law*, 1–39.

Higginson, S. (2014, April 14). Opinion: Public hearings a poor cousin to merit appeals.*Newcastle Herald*. Retrieved from www.theherald.com.au/story/2219279/ opinion-public-hearings-a-poor-cousin-to-merit-appeals/

Hillman, M. (2006). Situated justice in environmental decision-making: lessons from river management in Southeastern Australia. *Geoforum, 37*(5), 695–707.

Hilson, G. (2012), Corporate social responsibility in the extractive industries: experiences from developing countries. *Resources Policy, 37*(2), 131–137.

Holland, B. (2014). *Allocating the Earth: A Distributional Framework for Protecting Capabilities in Environmental Law and Policy*. Oxford: Oxford University Press.

Holloway, G. (2015, June 12). Industry warns of looming NSW gas shortage. *The Australian*. Retrieved from www.theaustralian.com.au/business/mining-energy/industry-warns-of-looming-nsw-gas-shortage/news-story/5de487a698f3af35c964adec5434c58c

Honneth, A. (1995). *The Struggle for Recognition: The Moral Grammar of Social Conflicts*. Cambridge, MA: MIT Press.

Hopey, D. (2016, April 18). Range Resources exec's well-site remarks drawing sharp criticism: Does Range avoid rich neighborhoods? *Pittsburgh Post-Gazette*. Retrieved from http://powersource.post-gazette.com/powersource/latest-oil-and-gas/2016/04/18/Executive-s-remark-about-shale-gas-well-sites-prompts-sharp-criticism-calls-for-review/stories/201604180027

Howard, T. M. (2017). 'Raising the bar': the role of institutional frameworks for community engagement in Australian natural resource governance. *Journal of Rural Studies, 49*, 78–91.

Huber, M. T., & Emel, J. (2009). Fixed minerals, scalar politics: the weight of scale in conflicts over the '1872 Mining Law' in the United States. *Environment and Planning A, 41*(2), 371–388.

Hudgins, A., & Poole A. (2014). Framing fracking: private property, common resources, and regimes of governance. *Journal of Political Ecology, 21*, 303–319.

Hunold, C., & Young, I. (1998). Justice, democracy, and hazardous siting. *Political Studies, 46*(1), 82–91.

Ikeme, J. (2003). Equity, environmental justice and sustainability: incomplete approaches in climate change politics. *Global Environmental Change, 13*(3), 195–206.

Independent Commission Against Corruption (ICAC). (2012). *Anti-corruption safeguards and the NSW planning system*. Retrieved from www.icac.nsw.gov.au/documents/preventing-corruption/cp-publications-guidelines/3867-anti-corruption-safeguards-and-the-nsw-planning-system-2012/file

Independent Expert Scientific Committee on Coal Seam Gas and Large Coal Mining Development (IESC). (2015). *Advice to decision maker on project: Watermark Coal Project (IESC 2015–066 and 2015–067)*. Retrieved from www.iesc.environment.gov.au/committee-advice/proposals/watermark-coal-project-new-development-project-advice-2015

IndexMundi. (2015). *Coal, Australian thermal coal Monthly Price – US dollars per metric ton*. Retrieved from www.indexmundi.com/commodities/?commodity=coal-australian& months=300

International Energy Agency. (2014). *World Energy Outlook 2014 Factsheet*. Retrieved from www.worldenergyoutlook.org/media/weowebsite/2014/WEO2014FactSheets.pdf

Jalbert, K. (2016, June 6). Environmental justice: failing PAs oil and gas communities. *FracTracker Alliance*. Retrieved from http://ft.maps.arcgis.com/apps/MapSeries/index. html?appid=149ae5ee334e4a03babf18c4c79feef9

Jarrell, M. L., Ozymy, J. & McGurrin, D. (2013). How to encourage conflict in the environmental decision-making process: imparting lessons from civic environmentalism to local policy-makers. *Local Environment, 18*(2), 184–200.

Jefferies, C. (2012). Unconventional bridges over troubled water: lessons to be learned from the Canadian oil sands as the United States moves to develop the natural gas of the Marcellus Shale play. *Energy Law Journal, 33*(1), 75–117.

Jentoft, S. (2013). Social justice in the context of fisheries – a governability challenge. In M. Bavinck, R. Chuenpagdee, S. Jentoft & J. Kooiman (Eds.), *Governability of Fisheries and Aquaculture* (pp. 45–65). Amsterdam: Springer.

Jessop, B. (1998). The rise of governance and the risks of failure: the case of economic development. *International Social Science Journal, 50*(155), 29 45.

Jessop, B. (2003). Governance and meta-governance: on reflexivity, requisite variety and requisite irony. In H. P. Bang (Ed.), *Governance as Social and Political Communication* (pp. 101–116). Manchester: Manchester University Press.

Jessup, B. (2013). Environmental justice as spatial and scalar justice: a regional waste facility or a local rubbish dump out of place? *International Journal of Sustainable Development Law and Policy, 9*(2), 70–107.

Jessup, B. (2015). Justice, recognition and environmental law: the Wielangta Forest conflict, Tasmania, Australia. *University of Tasmania Law Review, 34*(1), 5–33.

Jopson, D. (2011, March 1). Let them eat coal? Time to talk about food security. *Sydney Morning Herald*. Retrieved from www.smh.com.au/nsw/state-election-2011/let-them-eat-coal-time-to-talk-about-food-security-20110228-1bbsx.html#ixzz48xnVK6rv

Jordan, W., Clarke, P. & Flint, C. (2011). *Under the radar – how coal seam gas mining in the Pilliga is impacting matters of national environmental significance*. Prepared for the Wilderness Society, Nature Conservation Council of NSW and the Northern Inland Council for the Environment. Retrieved from https://cdn.wilderness.org.au/archive/files/Under%20 the%20Radar%20Eastern%20Star%20Gas%20EPBC%20Report%20email.pdf

Kelsey, T., Metcalf, A. & Salcedo, R. (2012). Marcellus Shale: land ownership, local voice, and the distribution of lease and royalty dollars. *Center for Economic and Community Development Research Paper Series*. Retrieved from http://aese.psu.edu/research/ centers/cecd/publications/marcellus/marcellus-shale-land-ownership-local-voice-and-the-distribution-of-lease-and-royalty-dollars

Kemp, D., Owen, J., Gotzmann, N. & Bond, C. (2011). Just relations and company–community conflict in mining. *Journal of Business Ethics, 101*(1), 93–109.

Kenyon-Slaney, H. (2013a, May 20). NSW court verdict on Rio's coalmine expansion a blow to jobs. *The Australian*. Retrieved from www.theaustralian.com.au/business/opinion/ nsw-court-verdict-on-rios-coalmine-expansion-a-blow-to-jobs/story-e6frg9if-1226646294204

Kenyon-Slaney, H. (2013b, August 14). Keep mining approvals out of court. *The Australian Financial Review*. Retrieved from www.afr.com/opinion/keep-mining-approvals-out-of-court-20130813-jhbj8

Kitze, R. A. (2013). Moving past preemption: enhancing the power of local governments over hydraulic fracturing. *Minnesota Law Review* 98, 385–418.

Kooiman, J., & Jentoft, S. (2009). Meta-governance: values, norms and principles, and the making of hard choices. *Public Administration*, 87(4), 818–836.

Kurtz, H. E. (2003). Scale frames and counter-scale frames: constructing the problem of environmental injustice. *Political Geography*, 22(8), 887–916.

Lagan, B. (2013, October 2). What happens with Rio Tinto, stays with Rio Tinto. *The Global Mail*.

Lake, R. W. (1996). Volunteers, NIMBYs, and environmental justice: dilemmas of democratic practice. *Antipode*, 28(2), 160–174.

Lin, C., & Lockwood, M. (2014). Assessing sense of place in natural settings: a mixed-method approach. *Journal of Environmental Planning and Management*, 57(10), 1444–1464.

Lipman, Z., & Stokes, R. (2008). The technocrat is back: environmental land-use planning reform in New South Wales. *Environmental and Planning Law Journal*, 25(5), 305–324.

Lock The Gate. (2016a, February 22). *Media release: Explosive documents reveal key policy changed for coal miners*. Retrieved from Lock the Gate website: www.lockthegate.org.au/acquisition_gipa_media

Lock The Gate. (2016b, May 20). *Media release: Lock the Gate challenges Minister Stokes to urgently restore legal appeal rights as Bulga residents withdraw challenge in Land & Environment Court*. Retrieved from Lock the Gate website: www.lockthegate.org.au/stokes_restore_merits_appeals_bulga

Longhurst, R. (2010). Semi-structured interviews and focus groups. In N. R. Clifford, S. French & G. Valentine (Eds.), *Key Methods in Geography* (pp. 117–132). Los Angeles: Sage.

Low, S., & Altman, I., (1992). Introduction. In I. Altman, & S. Low (Eds.), *Place Attachment* (pp. 1–12). New York: Plenum Press.

Lukes, S. (2005). *Power. A Radical View.* 2nd Edition. Basingstoke: Palgrave Macmillan.

Lynch, B. (1993). The garden and the sea: U.S. Latino environmental discourses and mainstream environmentalism. *Social Problems*, 40(1), 108–124.

McCarthy, J. (2014, February 19). Unjust mining laws slammed by former judges. *Sydney Morning Herald*. Retrieved from www.smh.com.au/nsw/unjust-mining-laws-slammed-by-former-judges-20140218-32zi0

McGee, B. (2009). The community referendum: participatory democracy and the right to free, prior and informed consent to development. *Berkeley Journal of International Law*, 27, 570–635.

McGrath, C. (2015, July 15). Shenhua mine: the federal government could have chosen farming over coal. *The Conversation*. Retrieved from https://theconversation.com/shenhua-mine-the-federal-government-could-have-chosen-farming-over-coal-44654

McHugh, B., Condon, M. & Herbert, L. (2014, 21 February). Santos signs agreement with NSW Government for Narrabri CSG project. *ABC Rural*. Retrieved from www.abc.net.au/news/2014-02-21/santos-nsw-government-signmou-for-narrabri-gas-project/5274804

MacKinnon, D. (2011). Reconstructing scale: towards a new scalar politics. *Progress in Human Geography*, 35(1), 21–36.

McManus, P., & Connor, L. (2013). What's mine is mine(d): contests over marginalisation of rural life in the Upper Hunter, NSW. *Rural Society*, 22(2), 166–183.

McManus, P., Albrecht, G. & Graham, R. (2014). Psychoterratic geographies of the Upper Hunter region, Australia. *Geoforum*, 51, 58–65.

McNellis, K. (2011). *Names in the news: Pennsylvania's Marcellus Shale Advisory Commission*. Helena, MT: National Institute on Money in State Politics.

Madden, J. (2011, June 28). Chinese land grab tests farming dynasties. *The Australian*. Retrieved from www.theaustralian.com.au/national-affairs/chinese-land-grab-tests-farming-dynasties/story-fn59niix-1226083089449

Maguire, L. A., & Lind, E. A. (2003). Public participation in environmental decisions: stakeholders, authorities and procedural justice. *International Journal of Global Environmental Issues*, 3(2), 133–148.

Marcellus Shale Advisory Commission. (2011). *Report*. Retrieved from www.dep.pa.gov/Business/Energy/OilandGasPrograms/OilandGasMgmt/MarcellusShale/Pages/default.aspx

Martin, A., Gross-Camp, N., Kebede, B., McGuire, S., & Munyarukaza, J. (2014). Whose environmental justice? Exploring local and global perspectives in a payments for ecosystem services scheme in Rwanda. *Geoforum*, 54, 167–177.

Martin, A., McGuire, S., & Sullivan, S. (2013). Global environmental justice and biodiversity conservation. *Geographical Journal*, 179(2), 122–131.

Martin, P., & Kennedy, A. (2016). Intra-national rivalries. In J. Gray, C. Holley & R. Rayfuse (Eds.), *Trans-jurisdictional Water Law and Governance* (pp. 101–119). Abingdon: Earthscan.

Martinez-Alier, J. (2001). Mining conflicts, environmental justice, and valuation. *Journal of Hazardous Materials*, 86(1), 153–170.

Martinez-Alier, J. (2012). Environmental justice and economic degrowth: an alliance between two movements. *Capitalism Nature Socialism*, 23(1), 51–73.

Mauriello, T., & Olson, L. (2011, March 13). Marcellus Shale Advisory Commission: out of balance? Shale panel long on contributors, drilling interests. *Pittsburgh Post-Gazette*. Retrieved from www.post-gazette.com/news/environment/2011/03/13/Marcellus-Shale-Advisory-Commission-out-of-balance/stories/201103130373

May, T. (2011). *Social Research*. Maidenhead: McGraw-Hill Education.

Measham, T. G. (2007). Building capacity for environmental management: local knowledge and rehabilitation on the Gippsland Red Gum Plains. *Australian Geographer*, 38(2), 145–159.

Mercer, A., de Rijke, K. & Dressler, W. (2014). Silences in the midst of the boom: coal seam gas, neoliberalizing discourse, and the future of regional Australia. *Journal of Political Ecology*, 21, 279–302.

Miller, D. (1999). *Principles of Social Justice*. Cambridge, MA: Harvard University Press.

Millner, F. (2011). Access to environmental justice. *Deakin Law Review*, 16(1), 189.

Mining and Petroleum Gateway Panel. (2014a). *Shenhua Watermark Coal Project Advisory Report*. Retrieved from https://majorprojects.affinitylive.com/public/a5c4a793b-6ba02545a334ebe30fc9b55/20140110_%20Gateway%20Panel%20Advice%20Shenhua%20Watermark%20Project.pdf

Mining and Petroleum Gateway Panel. (2014b). *Report by the Mining & Petroleum Gateway Panel to accompany a Conditional Gateway Certificate for the Caroona Coal Project*. Retrieved from https://majorprojects.affinitylive.com/public/b61f392aa6d9b9e6f768514140570cc5/Attachment%203%20-%20Mining%20&%20Petroleum%20Gateway%20Panel%20Report.pdf

Mondock, A. (2015). Shale I stay or shale I go? Pennsylvania's 'Marcellus Shale-size' of a debacle over fracking severance taxation. *Ave Maria Law Review*, 13, 119–141.

Morton, D. J. (2000). PEL 238 Gunnedah Basin coal seam methane. *New South Wales Mining and Exploration Quarterly (MINFO) No. 66*. St. Leonards, NSW: NSW Department of Mineral Resources.

Munro, S. (2012). *Rich Land Wasteland: How Coal is Killing Australia*. Sydney: Macmillan.

Myers, P. (2010, April 17). Coal brings cold comfort on the farm. *Sydney Morning Herald*. Retrieved from www.smh.com.au/business/coal-brings-cold-comfort-on-the-farm-20100416-skif.html#ixzz48nMIMqUX

Nadeau, R. L. (2006). *The Environmental Endgame: Mainstream Economics, Ecological Disaster, and Human Survival*. New Brunswick, NJ: Rutgers University Press.

Nader, J. (2013). A rose by any other name. *Bar News*, 5–6. Retrieved from www.austlii.edu.au/au/journals/NSWBarAssocNews/2013/3.pdf

Nader, J. (2014, January 8). Mining Acts need severe surgery. *The Land*. Retrieved from www.theland.com.au/story/3581975/mining-acts-need-severe-surgery/

Nagle, J. C. (2010). *Law's Environment: How the Law Shapes the Places We Live*. New Haven, CT: Yale University Press.

Namoi Valley Independent. (2013, November 14). Standing firm on fertile ground. *Namoi Valley Independent*. Retrieved from www.nvi.com.au/story/1908370/standing-firm-on-fertile-ground/?cs=372

Narrabri Shire Council. (2012). *Narrabri Shire economic profile*. Retrieved from www.narrabri.nsw.gov.au/files/uploaded/file/Economic%20Development/Narrabri_Shire_Economic_Profile.pdf

Neimanis, A., Castleden, H. & Rainham, D. (2012). Examining the place of ecological integrity in environmental justice: a systematic review. *Local Environment*, *17*(3), 349–367.

Newell, P. (2005). Race, class and the global politics of environmental inequality. *Global Environmental Politics*, 5(3), 70–94.

Nguyen, H. (2016, August 7). Bulga residents take the fight to Sydney in last bid to save their village. *Sydney Morning Herald*. Retrieved from www.smh.com.au/nsw/bulga-residents-take-the-fight-to-sydney-in-last-bid-to-save-their-village-20160807-gqmycl.html

Nicholls, S. (2011, October 2013). Angry farmers threaten to snub land-use conflicts group. *Sydney Morning Herald*. Retrieved from www.smh.com.au/nsw/angry-farmers-threaten-to-snub-landuse-conflicts-group-20111012-1ll8d.html#ixzz48yY1gjcl

Nicholls, S., & Hannam, P. (2016, March 7). Smaller penalties for CSG companies amid crack down on protesters. *Sydney Morning Herald*. Retrieved from www.smh.com.au/nsw/smaller-penalties-for-csg-companies-amid-crack-down-on-protesters-20160307-gncbkk.html#ixzz4AXRtYpgj

Nichols, L. (2016, July 18). Why good planning counts and why we should respect the courts. *Singleton Argus*. Retrieved from www.singletonargus.com.au/story/4038709/how-did-it-come-to-this-ignoring-justice/?cs=2038

Norris, E. H., Mitchell, P. B. & Hart, D. M. (1991). Vegetation changes in the Pilliga forests: a preliminary evaluation of the evidence. In A. Henderson-Sellers & A. J. Pitman (Eds.), *Vegetation and Climate Interactions in Semi-arid Regions* (pp. 209–218). Netherlands: Springer.

Northern Inland Council for the Environment. (2012). *The truth spills out: a case study of coal seam gas exploration in the Pilliga*. Retrieved from www.stoppilligacoalseamgas.com/wp-content/uploads/2011/12/The_Truth_Spills_Out_Final_May_2012_without_appendices.pdf

Northern Inland Council for the Environment, Colong Foundation for Wilderness, The Wilderness Society, & Nature Conservation Council of NSW. (2011). *Submission on: Eastern Star Gas Limited/Energy generation and supply (non-renewable)/25 km SW of Narrabri, 80km NNE of Coonabarabran/NSW/Narrabri Coal Seam Gas Field Development,*

Reference Number 2011/5914. Retrieved from www.stoppilligacoalseamgas.com/wp-content/uploads/2011/12/Federal_EPBC_Act_referral_submssion_April_2011.pdf

Northern Tier Regional Planning and Development Commission. (n.d.). *The NTRPDC Region*. Retrieved from www.northcrntier.org/region.php

Novotny, P. (1995). Where we live, work and play: reframing the cultural landscape of environmentalism in the environmental justice movement. *New Political Science, 17*(2), 61–79.

Noy, C. (2008). Sampling knowledge: the hermeneutics of snowball sampling in qualitative research. *International Journal of Social Research Methodology, 11*(4), 327–344.

NSW Government. (2014). *NSW gas plan*. Retrieved from https://www.nsw.gov.au/sites/default/files/miscellaneous/sc000218_nsw_gas_plan_announcement_web.pdf

NSW Government, Department of Industry. (n.d.). *MinView*. Retrieved from www.resourcesandenergy.nsw.gov.au/miners-and-explorers/geoscience-information/services/online-services/minview

NSW Government, Department of Industry Resources & Energy. (2014). *NSW coal industry profile*. Retrieved from www.resourcesandenergy.nsw.gov.au/investors/investment-opportunities/coal/coal-profile

NSW Government, Department of Industry Resources & Energy. (2016). *New South Wales – coal investment profile*. Retrieved from www.resourccsandenergy.nsw.gov.au/__data/assets/pdf_file/0005/584375/New-South-Wales-Coal-investment-profile-flyer.pdf

NSW Minerals Council. (2014). *Mining and farming thriving together*. Retrieved from www.worldclassminers.com.au/news/environment/mining-and-farming-thriving-together/

NSW National Parks and Wildlife Service. (n.d.). *Pilliga National Park*. Retrieved from www.nationalparks.nsw.gov.au/visit-a-park/parks/pilliga-national-park

Nussbaum, M. (2000). *Women and Human Development: The Capabilities Approach*. Cambridge: Cambridge University Press.

Nussbaum, M. (2006). *Frontiers of Justice: Disability, Nationality, Species Membership*. Cambridge, MA: Harvard University Press.

Nussbaum, M. C. (2011). *Creating Capabilities: The Human Development Approach*. Cambridge, MA: Harvard University Press.

O'Kane, M. (2014a). *Final Report of the Independent Review of Coal Seam Gas Activities in NSW*. Retrieved from NSW Government, Chief Scientist and Engineer website: www.chiefscientist.nsw.gov.au/__data/assets/pdf_file/0005/56912/140930-CSG-Final-Report.pdf

O'Kane, M. (2014b). *Independent Review of Coal Seam Gas Activities in NSW: study of regulatory compliance systems and processes for coal seam gas*. Retrieved from NSW Government, Chief Scientist and Engineer website: www.chiefscientist.nsw.gov.au/__data/assets/pdf_file/0006/56913/140930-Final-Compliance-Report.pdf

Ogneva-Himmelberger, Y., & Huang, L. (2015). Spatial distribution of unconventional gas wells and human populations in the Marcellus Shale in the United States: vulnerability analysis. *Applied Geography, 60*, 165–174.

Orgill, T. (2015). The perils of fast-tracking mining development: an examination of the *Mining SEPP* resource significance amendments. *Environment and Planning Law Journal, 32*(5), 486–501.

Osborne, N. (2015). Intersectionality and kyriarchy: a framework for approaching power and social justice in planning and climate change adaptation. *Planning Theory, 14*(2), 130–151.

Owens, K. (2012). Strategic regional land use plans: presenting the future for coal seam gas projects in New South Wales? *Environmental and Planning Law Journal, 29*(2), 113–128.

Özkaynak, B., Rodriguez-Labajos, B., Arsel, M., Avci, D., Carbonell, M. H., Chareyron, B., Chicaiza, G., Conde, M., Demaria, F., Finamore, R., Kohrs, B., Krishna, V. V., Mahongnao, M., Raeva, D., Singh, A. A., Slavov, T., Tkalec, T. & Yánez, I. (2012). *Mining conflicts around the world: common grounds from an environmental justice perspective. ISS Staff Group 4: Rural Development, Environment and Population.* Environmental Justice Organizations, Liabilities and Trade (EJOLT) Factsheet. Retrieved from http://hdl.handle.net/1765/38559.

Pacheco, E. (2015). It's a fracking conundrum: environmental justice and the battle to regulate hydraulic fracturing. *Ecology Law Quarterly, 42*(2), 373–395.

Paragreen, N., & Woodley, A. (2013). Social licence to operate and the coal seam gas industry: what can be learnt from already established mining operations? *Rural Society, 23*(1), 46–59.

Peake, T., Bell, S., Tame, T., Simpson, J. & Curran. T. (2002). *Warkworth Sands Woodland – an endangered ecological community distribution, ecological significance and conservation status. (Hunter Region Botanic Gardens Technical Paper.)* Retrieved from www.huntergardens.org.au/forms/Warkworth%20Sands%20Woodland.pdf

Penny, D., Williams, G., Gillespie, J. & Khem, R. (2016). 'Here be dragons': integrating scientific data and place-based observation for environmental management. *Applied Geography, 73*, 38–46.

Perkins, N. D. (2012). The fracturing of place: the regulation of Marcellus Shale development and the subordination of local experience. *Fordham Environmental Law Review, 23*, 44–79.

Perry, S. L. (2012). Environmental reviews and case studies: addressing the societal costs of unconventional oil and gas exploration and production: a framework for evaluating short-term, future, and cumulative risks and uncertainties of hydrofracking. *Environmental Practice, 14*(4), 352–365.

Perry, S. L. (2013). Using ethnography to monitor the community health implications of onshore unconventional oil and gas developments: examples from Pennsylvania's Marcellus Shale. *New Solutions: A Journal of Environmental and Occupational Health Policy, 23*(1), 33–53.

Philippopoulos-Mihalopoulos, A. (2010). Spatial justice: law and the geography of withdrawal. *International Journal of Law in Context, 6*(3), 201–216.

Philippopoulos-Mihalopoulos, A. (2011). Law's spatial turn: geography, justice and a certain fear of space. *Law, Culture and the Humanities, 7*(2), 187–202.

Philippopoulos-Mihalopoulos, A. (2014). *Spatial Justice: Body, Lawscape, Atmosphere.* New York: Routledge.

Pifer, R. H. (2010). Drake meets Marcellus: a review of Pennsylvania case law upon the sesquicentennial of the United States oil and gas industry. *Texas Journal of Oil, Gas, and Energy Law, 6*, 47–76.

Planning Assessment Commission. (2012). *Determination Report: Warkworth Extension Project (09_0202).* Retrieved from https://majorprojects.affinitylive.com/public/6fef9d4b5d37a581ced40ebe05fd5203/Project%20Approval.pdf

Planning Assessment Commission. (2014a). *Watermark Coal Project: Review Report.* Retrieved from https://majorprojects.affinitylive.com/public/bfbc8bbf7617082e17084b9a2654d0a8/2.%20Watermark%20Coal%20Project%20-%20PAC%20Review%20Report.pdf

Planning Assessment Commission. (2014b). *Determinations of the Narrabri Coal Seam Gas Utilisation Project (MP07_0023 MOD 3); the Bibblewindi Gas Exploration Pilot Expansion Project SSD 5934; and the Dewhurst Gas Exploration Pilot Expansion Project SSD 6038.* Retrieved from https://majorprojects.affinitylive.com/public/

c0129418f776474ddc969b8e2a996673/Bibblewindi%20Gas%20Exploration%20
Pilot%20Expansion_PAC%20Determination%20Report.pdf

Planning Assessment Commission. (2014c). *Determination Report: Warkworth Coal Mine Modification 6, Singleton LGA*. Retrieved from http://dev.pac.nsw.gov.au/resources/pac/media/files/pac/projects/2013/12/warkworth-coal-project-mod-6/pac-determination/determination-report-290114pdf.pdf

Planning Assessment Commission. (2015a). *Warkworth Continuation Project: Review Report*. Retrieved from https://majorprojects.affinitylive.com/public/223c09592b221 16450e639b6ff2daa38/Warkworth%20Continuation%20Project%20-%20PAC%20Review%20Report.pdf

Planning Assessment Commission. (2015b). *Mount Thorley Continuation Project: Review Report*. Retrieved from https://majorprojects.affinitylive.com/public/44c9d0e3ab9760d cb05a4a1af9c88ee2/Mt%20Thorley%20Continuation%20Project%20-%20PAC%20Review%20Report.pdf

Planning Assessment Commission. (2015c). *Warkworth Continuation Project: Second Review Report*. Retrieved from https://majorprojects.affinitylive.com/public/e2fe9189476109eb28ff45ecb0549e3c/04.%20Warkworth%20Continuation%20Project%20-%20PAC%20Second%20Review%20Report.pdf

Planning Assessment Commission. (2015d). *Mount Thorley Continuation Project: Second Review Report*. Retrieved from www.pac.nsw.gov.au/resources/pac/media/files/pac/projects/2015/08/mount-thorley-continuation-project-second-review/pacs-second-review-report/second-mt-thorley-review-reportpdf.pdf

Planning Assessment Commission. (2015e). *Determination Report: Warkworth Continuation Project SSD6464*. Retrieved from www.pac.nsw.gov.au/resources/pac/media/files/pac/projects/2015/05/warkworth-continuation-project—determination/determination-report/1-warkworthdeterminationreportpdf.pdf

Planning Assessment Commission. (2015f). *Mount Thorley Continuation Project (SSD 6465) Determination Report*. Retrieved from www.pac.nsw.gov.au/resources/pac/media/files/pac/projects/2015/05/mt-thorley-continuation-project—determination/determination-report/1-mtthorleydeterminationreportpdf.pdf

Planning Assessment Commission. (2015g). *NSW Planning Assessment Commission Determination Report: Watermark Coal Project, Gunnedah LGA*. Retrieved from https://majorprojects.affinitylive.com/public/3013970e3a05fb74a629ac7a8ad8a73e/Watermark%20Coal%20Project%20-%20PAC%20Determination%20Report.pdf

Plumwood, V. (2001). Nature as agency and the prospects for a progressive naturalism. *Capitalism Nature Socialism, 12*(4), 3–32.

Preston, B. J. (2015a). The adequacy of the law in satisfying society's expectations for major projects. *Environment and Planning Law Journal, 32*, 182–201.

Preston, B. J. (2015b). The effectiveness of the law in providing access to environmental justice: an introduction. In P. Martin, S. Z. Bigdeli, T. Daya-Winterbottom, W. du Plessis & A. Kennedy (Eds.), *The Search for Environmental Justice* (pp. 23–43). Cheltenham: Edward Elgar.

Pulido, L. (1996). *Environmentalism and Economic Justice: Two Chicano Struggles in the Southwest*. Tucson: University of Arizona Press.

Queensland Department of Natural Resources and Mines. (2012). *Great Artesian Basin Water Resource Plan – Five Year Review*. Retrieved from https://www.dnrm.qld.gov.au/__data/assets/pdf_file/0018/106029/5year-wrp-review.pdf

Radow, E. N. (2010). Citizen David tames gas Goliaths on the Marcellus Shale stage: citizen action as a form of dispute prevention in the internet age. *Cardozo Journal of Conflict Resolution, 12*, 373–410.

Rahm, D. (2011). Regulating hydraulic fracturing in shale gas plays: the case of Texas. *Energy Policy, 39*(29), 74–81.

Rapley, T. J. (2001). The art(fulness) of open-ended interviewing: some considerations on analysing interviews. *Qualitative Research, 1*(3), 303–323.

Rapley, T. J. (2007). *Doing Conversation, Discourse and Document Analysis.* Los Angeles: Sage.

Rawls, J. (1971). *A Theory of Justice.* Cambridge, MA: Harvard University Press.

Razaghi, T. (2016, November 4). Aboriginal elders speak out in support of heritage protection on proposed Shenhua mine site. *ABC New England Online.* Retrieved from www.abc.net.au/news/2016-11-04/emotional-consultation-aboriginal-heritage-shenhua-mine/7996884

Razaghi, T., Fuller, K. & Thomas, K. (2016, September 9). Investigation into Aboriginal heritage on proposed Shenhua mine site launched. *ABC New England Online.* Retrieved from www.abc.net.au/news/2016-09-09/invesigation-launched-aboriginal-heritage-shenhua-mine-site/7829648

Reed, M. S. (2008). Stakeholder participation for environmental management: a literature review. *Biological Conservation, 141*(10), 2417–2431.

Richardson, B. J., & Razzaque, J. (2006). Public participation in environmental decision-making. In B. J. Richardson & S. Wood (Eds.), *Environmental Law for Sustainability* (pp. 165–194). Oxford: Hart Publishing.

RioTinto. (2014, March 20). Mount Thorley Warkworth seeks long term future for 1300 workers. Retrieved from www.riotinto.com/media/media-releases-237_10381.aspx

RioTinto. (n.d.). *Mount Thorley Warkworth.* Retrieved from www.riotinto.com/copperand-coal/mount-thorley-warkworth-10427.aspx

Robeyns, I. (2003). Is Nancy Fraser's critique of theories of distributive justice justified? *Constellations, 10*(4), 538–554.

Robeyns, I. (2005). The capability approach: a theoretical survey. *Journal of Human Development, 6*(1), 93–117.

Robeyns, I. (2006). The capability approach in practice. *Journal of Political Philosophy, 14*(3), 351–376.

Robins, B., & Cubby, B. (2010, April 21). Court ruling blocking mine access sidestepped. *Sydney Morning Herald.* Retrieved from www.smh.com.au/environment/court-ruling-blocking-mine-access-sidestepped-20100420-sru7.html#ixzz48n1OEOfs

Robinson, J. (2013, January 2). *The evolution of a word – the case of 'Clayton's'.* Retrieved from Ozwords – Australian National Dictionary Centre website: http://ozwords.org/?p=3240

Roesler, S. (2011). Addressing environmental injustices: a capability approach to rulemaking. *West Virginia Law Review, 114,* 49–107.

Rolls, E. (1981). *A million wild acres: 200 years of man and an Australian forest.* Melbourne: Nelson.

Rose, D. B. (1996). *Nourishing Terrains: Australian Aboriginal Views of Landscape and Wilderness.* Canberra: Australian Heritage Commission.

Rosenfeld, S. (2012, March 7). Fracking democracy: Why Pennsylvania's Act 13 may be the nation's worst corporate giveaway. *Alternet.* Retrieved from www.alternet.org/story/154459/fracking_democracy%3A_why_pennsylvania's_act_13_may_be_the_nation's_worst_corporate_giveaway

Rubinkam, M. (2011, April 13). Pennsylvania is approving gas drilling permits with scant review. *Associated Press.* Retrieved from http://usatoday30.usatoday.com/money/industries/energy/2011-04-13-pa-gas-drilling-permits.htm

Rutovitz, J., Harris, S., Kuruppu, N. & Dunstan, C. (2011). *Drilling down: coal seam gas – a background paper.* Prepared for the City of Sydney by the Institute for Sustainable

Futures and the University of Technology, Sydney. Retrieved from https://opus.lib.uts.edu.au/handle/10453/31334

Sangaramoorthy, T., Jamison, A. M., Boyle, M. D., Payne-Sturges, D. C., Sapkota, A., Milton, D. K. & Wilson, S. M. (2016). Place-based perceptions of the impacts of fracking along the Marcellus Shale. *Social Science & Medicine*, *151*, 27–37.

Santos. (2011, November 17). *Media release: Santos completes acquisition of Eastern Star Gas*. Retrieved from Santos website: https://www.santos.com/media-centre/announcements/santos-completes-acquisition-of-eastern-star-gas/

Santos. (2012, February 22). *Media release: Review of Eastern Star Gas operations*. Retrieved from Santos website: https://www.santos.com/media-centre/announcements/review-of-eastern-star-gas-operations-1/

Santos. (2013a). *Leewood produced water and brine management ponds: review of environmental factors*. Prepared by RPS Australia Ltd. Retrieved from http://digsopen.minerals.nsw.gov.au/?rin=R00070569

Santos. (2013b, June 28). *Media release: Santos seeks approval for Pilliga exploration program*. Retrieved from: https://www.santos.com/media-centre/announcements/santos-seeks-approval-for-pilliga-exploration-program-1/

Santos. (2014, 21 February). *Media release: Santos and NSW Government reach agreement to secure affordable supply of natural gas from the Narrabri Gas Project*. Retrieved from: https://www.santos.com/media-centre/announcements/santos-and-nsw-government-reach-agreement-to-secure-affordable-supply-of-natural-gas-from-the-narrabri-gas-project/

Santos. (2015a). *The Narrabri Gas Project: Overview*. Retrieved from https://narrabrigasproject.com.au/uploads/2015/04/FACT-SHEET-Narrabri-Gas-May-2015-FINAL-Web-other.pdf

Santos. (2015b). *Leewood Produced Water Treatment and Beneficial Reuse Project: review of environmental factors*. Prepared by RPS Australia East Pty Ltd. Retrieved from http://digsopen.minerals.nsw.gov.au/?rin=R00070569

Santos. (n.d). *Narrabri Gas Project*. Retrieved from https://www.santos.com/what-we-do/activities/new-south-wales/gunnedah-basin/narrabri-gas-project/

Schafft, K., & Biddle, C. (2015). Opportunity, ambivalence, and youth perspectives on community change in Pennsylvania's Marcellus Shale region. *Human Organization*, *74*(1), 74.

Schattschneider, E. (1960). *The Semi-Sovereign People: A Realist's View of Democracy in America*. New York: Holt, Rinehart and Winston.

Schlosberg, D. (2007). *Defining Environmental Justice: Theories, Movements, and Nature*. Oxford: Oxford University Press.

Schlosberg, D. (2012). Climate justice and capabilities: a framework for adaptation policy. *Ethics & International Affairs*, *26*(04), 445–461.

Schlosberg, D. (2013). Theorising environmental justice: the expanding sphere of a discourse. *Environmental Politics*, *22*(1), 37–55.

Schlosberg, D., & Carruthers, D. (2010). Indigenous struggles, environmental justice, and community capabilities. *Global Environmental Politics*, *10*(4), 12–35.

Sen, A. (1993). Capability and well-being. In M. Nussbaum & A. Sen (Eds.), *The Quality of Life* (pp. 30–53). Oxford: Clarendon Press.

Sen, A. (1999). *Development as Freedom*. Oxford: Oxford University Press.

Sen, A. (2005). Human rights and capabilities. *Journal of Human Development*, *6*(2), 151–166.

Sharma-Wallace, L. (2013). 'But you never know in these kind of things': contingent factors for environmental justice at the Beare Wetland, Scarborough, Canada. *Environmental Justice*, *6*(6), 206–212.

Sharp, E., & Curtis, A. (2012). *Groundwater management in the Namoi: a social perspective. Report No. 67.* Retrieved from Institute for Land, Water and Society, Charles Sturt University website: https://www.csu.edu.au/__data/assets/pdf_file/0005/702662/67_Namoi.pdf

Shenhua Australia Holdings Pty Ltd. (2016). *China Shenhua Energy Company Limited: announcement of annual results for the year ended 31 December 2015.* Retrieved from www.hkexnews.hk/listedco/listconews/sehk/2016/0324/LTN201603241361.pdf

Shenhua Australia Holdings Pty Ltd. (n.d.). *Watermark Project.* Retrieved from www.shenhuawatermark.com/shaus/1382705831568/The_Project.shtml

Sherval, M., & Graham, N. (2013). Missing the connection: how the strategic regional land use policy fragments landscapes and communities in NSW. *Alternative Law Journal, 38*(3), 176–180.

Sherval, M., & Hardiman, K. (2014). Competing perceptions of the rural idyll: responses to threats from coal seam gas development in Gloucester, NSW, Australia. *Australian Geographer, 45*(2), 185–203.

Shrader-Frechette, K. (2002). *Environmental Justice: Creating Equality, Reclaiming Democracy.* Oxford: Oxford University Press.

Sica, C. E. (2015). Stacked scale frames: building hegemony for fracking across scales. *Area, 47*(4), 443–450.

Simcock, N. (2014). Exploring how stakeholders in two community wind projects use a 'those affected' principle to evaluate the fairness of each project's spatial boundary. *Local Environment, 19*(3), 241–258.

Simonelli, J. (2014). Home rule and natural gas development in New York: civil fracking rights. *Journal of Political Ecology, 21*(1), 258–278.

Simpson, H., de Loë, R. & Andrey, J. (2015). Vernacular knowledge and water management – towards the integration of expert science and local knowledge in Ontario, Canada. *Water Alternatives, 8*(3), 352–372.

Simson, F. (2014, December 19). Watermark: a failed planning watershed? *The Land.* Retrieved from www.theland.com.au/story/3373231/watermark-a-failed-planning-watershed/

Skitka, L., Winquist, J. & Hutchinson, S. (2003). Are outcome fairness and outcome favorability distinguishable psychological constructs? A meta-analytic review. *Social Justice Research, 16*(4), 309–341.

Smerdon, B., Ransley, T., Radke, B. & Kellett, J. (2012). *Water resource assessment for the Great Artesian Basin. A report to the Australian Government from the CSIRO Great Artesian Basin Water Resource Assessment.* CSIRO: Water for a Healthy Country Flagship. Retrieved from CSIRO website: https://publications.csiro.au/rpr/download?pid=csiro:EP132685&dsid=DS3

Smith, J. M. (2011). The prodigal son returns: oil and gas drillers return to Pennsylvania with a vengeance – Are municipalities prepared? *Duquesne Law Review, 49*, 1–34.

Smith, M. F., & Ferguson, D. P. (2013). 'Fracking democracy': issue management and locus of policy decision-making in the Marcellus Shale gas drilling debate. *Public Relations Review, 39*(4), 377–386.

Smith, P. D., & McDonough, M. H. (2001). Beyond public participation: fairness in natural resource decision making. *Society & Natural Resources, 14*(3), 239–249.

SoilFutures Consulting Pty Ltd. (2015). *Great Artesian Basin Recharge Systems and Extent of Petroleum and Gas Leases: Second Edition, with response to Ministerial Review.* Prepared for the Artesian Bore Water Users Association.

Sovacool, B. K. (2014). Cornucopia or curse? Reviewing the costs and benefits of shale gas hydraulic fracturing (fracking). *Renewable and Sustainable Energy Reviews, 37*, 249–264.

Sovacool, B. K., & Dworkin, M. H. (2015). Energy justice: conceptual insights and practical applications. *Applied Energy, 142*, 435–444.

Spence, D. B. (2012). Federalism, regulatory lags, and the political economy of energy production. *University of Pennsylvania Law Review, 161*, 431–508.

Spence, D. B. (2014). The political economy of local vetoes. *Texas Law Review, 93*, 351–413.

Stake, R. E. (1995). *The Art of Case Study Research*. Los Angeles: Sage.

Stares, D., McElfish, J. & Ubinger Jr, J. (2016). Sustainability and community responses to local impacts. In J. C. Dernbach & J. R. May (Eds.), *Shale Gas and the Future of Energy: Law and Policy for Sustainability* (pp. 101–125). Cheltenham: Edward Elgar.

Stoner, A., MP (Deputy Premier of NSW), & Roberts, A., MP (Minister for Resources and Energy). (2014, February 21). *Media release: Memorandum of Understanding on Narrabri Gas Project*. Retrieved from www.resourcesandenergy.nsw.gov.au/__data/assets/ pdf_file/0010/531865/140214-MOU-narrabri-project.pdf

Sturmer, J., & Armitage, R. (2015, May 21). Santos' Narrabri coal seam gas project under scrutiny from government departments. *ABC News Online*. Retrieved from www.abc. net.au/news/2015-05-21/narrabri-coal-seam-gas-project-hits-hurdle/6488504

Syme, G., & Nancarrow, B. (2008). Justice and the allocation of benefits from water. *Social Alternatives, 27*(3), 21–25.

Syme, G., & Nancarrow, B. (2012). Justice and the allocation of natural resources: current concepts and future directions. In S. Clayton (Ed.), *The Oxford Handbook of Environmental and Conservation Psychology* (pp. 93–112). Oxford: Oxford University Press.

Syme, G. J., Nancarrow, B. E. & McCreddin, J. A. (1999). Defining the components of fairness in the allocation of water to environmental and human uses. *Journal of Environmental Management, 57*(1), 51–70.

Szasz, A., & Meuser, M. (1997). Public participation in the cleanup of contaminated military facilities: democratization or anticipatory cooptation? *International Journal of Contemporary Sociology, 34*(1), 211–234.

Tasker, S. (2013, May 13). Minister joins Rio Tinto in Warkworth mine appeal. *The Australian*. Retrieved from www.theaustralian.com.au/national-affairs/state-politics/ minister-joins-rio-tinto-in-warkworth-mine-appeal/story-e6frgczx-1226640706774

Taylor, D. E. (1997). Women of color, environmental justice, and ecofeminism. In K. Warren & N. Erkal (Eds.), *Ecofeminism: Women, Culture, Nature* (pp. 38–81). Bloomington: Indiana University Press.

Thompson, H. (2013, September 29). State government planning changes could put communities at risk. *ABC News*. Retrieved from http://blogs.abc.net.au/nsw/2013/07/state-government-planning-changes-could-put-communities-at-risk.html?site=newcastle&p rogram=newcastle_mornings

Thomson, P. (2010, March 24). Caroona blockade to be lifted today. *Australian Dairy Farmer*. Retrieved from http://adf.farmonline.com.au/news/state/agribusiness/general-news/caroona-blockade-to-be-lifted-today/1783663.aspx

Tschakert, P. (2009). Digging deep for justice: a radical re-imagination of the artisanal gold mining sector in Ghana. *Antipode, 41*(4), 706–740.

Turton, D. J. (2015). Unconventional gas in Australia: towards a legal geography. *Geographical Research, 53*(1), 53–67.

Tyler, R. (2000). Social justice: outcome and procedure. *International Journal of Psychology, 35*(2), 117–125.

Tyler, R., & Blader, S. (2003). The group engagement model: procedural justice, social identity, and cooperative behaviour. *Personality and Social Psychology Review, 7*(4), 349–361.

United Church of Christ Commission for Racial Justice. (1987). *Toxic Wastes and Race in the United States: A National Report on the Racial and Socio-Economic Characteristics of Communities with Hazardous Waste Sites*. New York: United Church of Christ.

Upstream Petroleum Consulting Services. (2000). *Assessment of Hydrocarbon Potential: Southern Brigalow Bioregion, Pilliga, New South Wales*. Prepared for the New South Wales Department of Mineral Resources. Retrieved from www.epa.nsw.gov.au/resources/forestagreements/pilligagastext.pdf

Urkidi, L., & Walter, M. (2011). Dimensions of environmental justice in anti-gold mining movements in Latin America. *Geoforum, 42*(6), 683–695.

Van Lieshout, M., Dewulf, A., Aarts, N. & Termeer, C. (2011). Do scale frames matter? Scale frame mismatches in the decision making process of a 'mega farm' in a small Dutch village. *Ecology and Society, 16*(1), 38. Retrieved from www.ecologyandsociety.org/vol16/iss1/art38

Van Lieshout, M., Dewulf, A., Aarts, N. & Termeer, C. (2014). The power to frame the scale? Analysing scalar politics over, in and of a deliberative governance process. *Journal of Environmental Policy & Planning*, 1–24.

Van Wagner, E. (2016a). Law's rurality: land use law and the shaping of people–place relations in rural Ontario. *Journal of Rural Studies, 47*, 311–325.

Van Wagner, E. (2016b). Law's ecological relations: the legal structure of people–place relations in Ontario's aggregate extraction conflicts. *Projections: The MIT Journal of Planning, 12*, 35–70.

Vanclay, F. (2003). International principles for social impact assessment. *Impact Assessment & Project Appraisal, 21*(1), 5–11.

Vanclay, F., Esteves, A. M., Aucamp, I. & Franks, D. M. (2015). *Social Impact Assessment: Guidance for assessing and managing the social impacts of projects*. Fargo, ND: International Association for Impact Assessment.

Voss, M., & Greenspan, E. (2012). Community consent index: oil, gas and mining company public positions on free, prior, and informed consent (FPIC). Oxfam America. Retrieved from www.oxfamamerica.org/explore/research-publications/community-consent-index/

Wagstaff, W. D. (2012). Fractured Pennsylvania: an analysis of hydraulic fracturing, municipal ordinances, and the Pennsylvania Oil and Gas Act. *New York University Environmental Law Journal, 20*, 327–361.

Wahlquist, A. (2010, August 28). Coal pays top dollar for foodbowl. *The Australian*. Retrieved from www.theaustralian.com.au/national-affairs/chinese-miner-shenhua-watermark-coal-pays-top-dollar-for-foodbowl/story-fn59niix-1225911097609

Walker, B. (2014). *Examination of the Land Access Arbitration Framework: Mining Act 1992 and Petroleum (Onshore) Act 1991*. Retrieved from www.resourcesandenergy.nsw.gov.au/__data/assets/pdf_file/0018/527112/Brett-Walker-Examination-of-the-Land-Access-Arbitration-Framework.pdf

Walker, G. (2009). Beyond distribution and proximity: exploring the multiple spatialities of environmental justice. *Antipode, 41*(4), 614–636.

Walker, G. (2012). *Environmental Justice: Concepts, Evidence and Politics*. London: Routledge.

Walker, G., & Bulkeley, H. (2006). Geographies of environmental justice. *Geoforum, 37*(5), 655–659.

Water Research Laboratory. (2014). *Review of Watermark Coal Groundwater Impact Assessment*. Presentation to the NSW PAC by Mr Doug Anderson on behalf of the NSW Irrigators Council.

Water Research Laboratory. (2015). *Independent Peer Review of Watermark Coal Project Groundwater Modelling Report*. Retrieved from www.wrl.unsw.edu.au/independent-peer-review-of-watermark-coal-project-groundwater-modelling-report

Weiland, P. S. (1999). Preemption of local efforts to protect the environment: implications for local government officials. *Virginia Environmental Law Journal, 18*, 467–506.

Wen, P. (2013, May 1). Mine rejection will cost jobs, Rio Tinto warns. *Sydney Morning Herald*. Retrieved from www.smh.com.au/business/mine-rejection-will-cost-jobs-rio-tinto-warns-20130430-2ir1q.html

Wenz, P. (1988). *Environmental Justice*. Albany: State University of New York Press.

Western Conservation Alliance, National Parks Association of NSW, Nature Conservation Council of NSW, Total Environment Centre, & Wilderness Society. (2001, April 12). *Media release: Narrabri Gas Field – an environmental disaster*. Retrieved from http://dazed.org/npa/press/20010412Narabri.htm

Westra, L. (1994). *An Environmental Proposal for Ethics: The Principle of Integrity*. Lanham, MD: Rowman & Littlefield.

White, R. (2013). Resource extraction leaves something behind: environmental justice and mining. *International Journal for Crime, Justice and Social Democracy, 2*(1), 50–64.

Whiting, F. (2016, June 11). George Bender: a bitter harvest. *The Courier Mail*. Retrieved from www.couriermail.com.au/news/queensland/george-bender-a-bitter-harvest/news-story/71f04e163f143a1cc699a90b9356b944

Wilber, T. (2012). *Under the Surface: Fracking, Fortunes, and the Fate of Marcellus Shale*. Ithaca, NY: Cornell University Press.

Willig, C. (2003). Discourse analysis. In J. A. Smith (Ed.), *Qualitative Psychology: A Practical Guide to Research Methods* (pp. 159–183). Los Angeles: Sage.

Willow, A. J. (2016). Wells and well-being: neoliberalism and holistic sustainability in the shale energy debate. *Local Environment, 21*(6), 768–788.

Willow, A. J., Zak, R., Vilaplana, D. & Sheeley, D. (2014). The contested landscape of unconventional energy development: a report from Ohio's shale gas country. *Journal of Environmental Studies and Sciences, 4*(1), 56–64.

Winestock, G., & Riordan, P. (2015, July 16). Greg Hunt announces Shenhua mine review on radio. *Australian Financial Review*. Retrieved from www.afr.com/business/mining/coal/greg-hunt-announces-shenhua-mine-review-on-radio-20150716-gidoxu

Wiseman, H. J. (2010). Regulatory adaptation in fractured Appalachia. *Villanova Environmental Law Journal, 21*, 229–292.

Wiseman, H. J. (2014). Governing fracking from the ground up. *Texas Law Review, 93*, 29–46.

WorleyParsons. (2013). *Groundwater risks associated with Coal Seam Gas Development in the Surat and Southern Bowen Basins – Final Report*. Prepared for the Australian Department of Natural Resources and Mines. Retrieved from https://www.dnrm.qld.gov.au/__data/assets/pdf_file/0013/106015/act-5-groundwater-risks-report-text.pdf

Wrenn, D. H., Kelsey, T. W. & Jaenicke, E. C. (2015). Resident vs. nonresident employment associated with Marcellus Shale development. *Agricultural and Resource Economics Review, 44*(2), 1–19.

Yin, R. K. (2009). *Case Study Research: Design and Methods*. Los Angeles: Sage.

Young, I. (1990). *Justice and the Politics of Difference*. Princeton, NJ: Princeton University Press.

Index

Milton Keynes UK
Ingram Content Group UK Ltd.
UKHW040102071024
449327UK00019B/753

9 780367 335311